**Standard Grade | Credit**

# Biology

| | |
|---|---|
| Credit Level 2001 | ✓ |
| Credit Level 2002 | ✓ |
| Credit Level 2003 | ✓ |
| Credit Level 2004 | ✓ |
| Credit Level 2005 | ✓ |

Leckie×Leckie

© **Scottish Qualifications Authority**

All rights reserved. Copying prohibited. No part of this publication may be reproduced, stored in a retrieval system, or transmitted in any form or by any means, electronic, mechanical, photocopying, recording or otherwise.

First exam published in 2001.
Published by Leckie & Leckie, 8 Whitehill Terrace, St. Andrews, Scotland KY16 8RN  tel: 01334 475656  fax: 01334 477392
enquiries@leckieandleckie.co.uk  www.leckieandleckie.co.uk

ISBN 1-84372-290-9

A CIP Catalogue record for this book is available from the British Library.

Printed in Scotland by Scotprint.

Leckie & Leckie is a division of Granada Learning Limited, part of ITV plc.

# Acknowledgements

Leckie & Leckie is grateful to the copyright holders, as credited at the back of the book, for permission to use their material.
Every effort has been made to trace the copyright holders and to obtain their permission to use their copyright material.
Leckie & Leckie will gladly receive information enabling them to rectify any error or omission in subsequent editions.

2001 | Credit

FOR OFFICIAL USE

| | | | | | |
|---|---|---|---|---|---|

C

KU | PS

Total Marks

**0300/402**

NATIONAL QUALIFICATIONS 2001

MONDAY, 21 MAY 10.50 AM – 12.20 PM

**BIOLOGY STANDARD GRADE** Credit Level

**Fill in these boxes and read what is printed below.**

Full name of centre

Town

Forename(s)

Surname

Date of birth
Day Month Year    Scottish candidate number    Number of seat

1  All questions should be attempted.

2  The questions may be answered in any order but all answers are to be written in the spaces provided in this answer book, and must be written clearly and legibly in ink.

3  Rough work, if any should be necessary, as well as the fair copy, is to be written in this book. Additional spaces for answers and for rough work will be found at the end of the book. Rough work should be scored through when the fair copy has been written.

4  Before leaving the examination room you must give this book to the invigilator. If you do not, you may lose all the marks for this paper.

1. A garden compost heap was marked off into five zones as shown below.

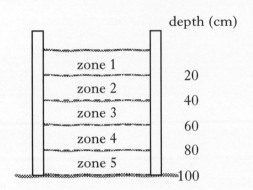

Three samples were removed from each zone and the average biomass of animals was calculated.

The results are shown in the table below.

| Animal | Average biomass of animals (mg/l) | | | | |
|---|---|---|---|---|---|
| | Zone 1 | Zone 2 | Zone 3 | Zone 4 | Zone 5 |
| Earthworms | 300 | 114 | 96 | 51 | 36 |
| Slugs | 258 | 63 | 54 | 0 | 0 |
| Woodlice | 204 | 87 | 75 | 33 | 6 |
| Centipedes | 9 | 18 | 18 | 15 | 12 |
| Insects | 6 | 6 | 3 | 0 | 0 |
| Mites | 12 | 12 | 6 | 3 | 3 |
| Total | 789 | 300 | 252 | 102 | 57 |

(a) Which animal contributes most biomass to the whole compost heap?

_____

(b) What percentage of the total animal biomass of the compost heap is composed of insects?

*Space for calculation*

_____ %

(c) Why were three samples taken from each zone?

_____

(d) What trend is shown by the total animal biomass as the depth increases?

_____

**2.** The Biochemical Oxygen Demand (BOD) indicates the level of the organic matter in water samples. The more organic matter present, the higher the BOD.

The diagram shows four sites on a river where water was sampled and the BOD measured. The sewage treatment plant was not working and untreated sewage was flowing into the river.

The following BODs were obtained: 8, 30, 93 and 126.

(a) Complete the diagram by writing the correct BOD at each sample site.

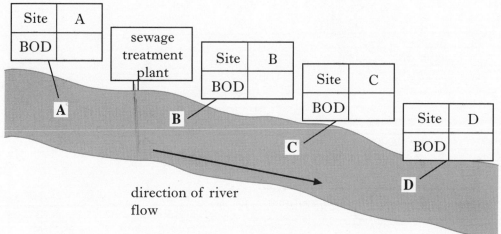

direction of river flow

(b) The BOD measures how much oxygen is used by microorganisms in the water.

Explain why a high organic matter content in the water will result in a high BOD.

_____

_____

(c) What term is used for a type of organism whose presence or absence gives information about pollution levels?

_____

**3.** The diagram below represents part of an investigation into heat production by germinating pea seeds.

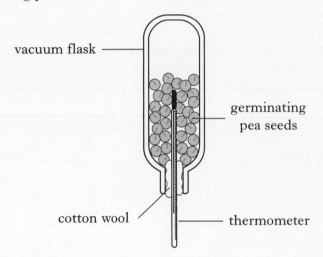

The temperature inside the flask was recorded for 72 hours.
The results are shown below.

| Time (hours) | Temperature (°C) |
|---|---|
| 0 | 18 |
| 12 | 26 |
| 24 | 40 |
| 36 | 48 |
| 48 | 50 |
| 60 | 52 |
| 72 | 54 |

(a) Calculate the average temperature rise per hour.
*Space for calculation*

Average temperature rise _____ °C per hour

3. (continued)

(b) On the grid below, complete the Y-axis and plot a **line graph** of the results.

(Additional graph paper, if required, will be found on page 27.)

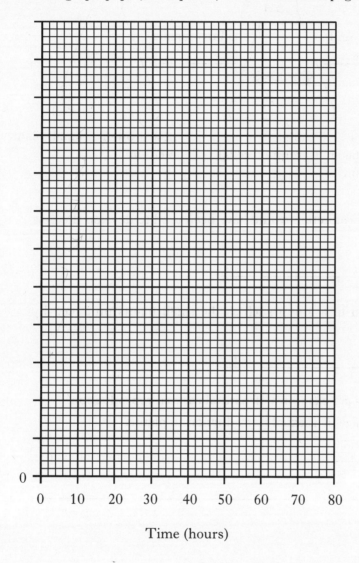

Time (hours)

2

(c) It was concluded that germinating seeds release heat energy.

Describe **one** way in which a control experiment should be kept the same as the first experiment, and **one** way in which it should differ, to make this conclusion valid.

Kept the same _____

Made different _____

2

[Turn over

DO NOT WRITE IN THIS MARGIN

| Marks | KU | PS |
|---|---|---|

**4.** After a new variety of rose has been developed, large numbers are produced for sale by artificial propagation techniques involving asexual reproduction.

The diagram shows artificial propagation by tissue culture.

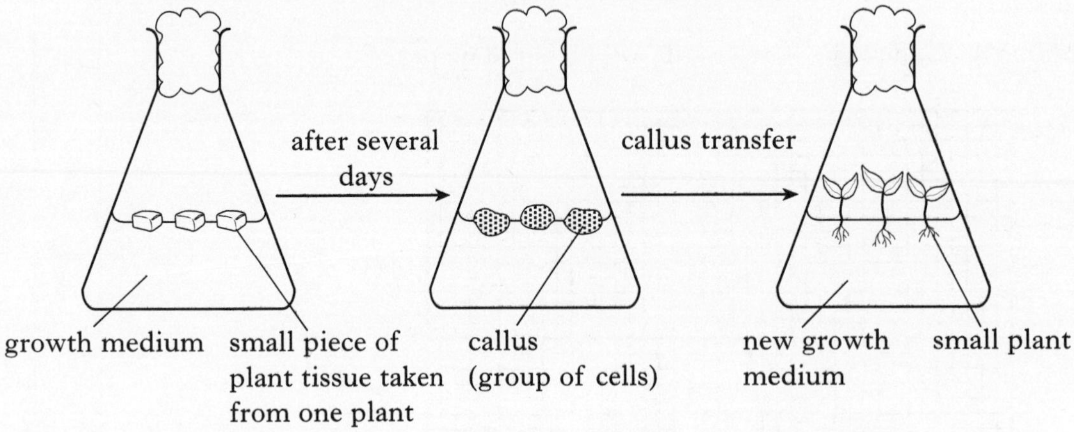


(*a*) What method of reproduction would have been used to develop the new variety of rose?

________________________________ **1**

(*b*) What name is given to a group such as the small plants produced by tissue culture?

________________________________ **1**

(*c*) Runners and tubers are examples of natural asexual reproduction.

Describe an advantage of asexual reproduction to plants.

________________________________________________________

________________________________________________________ **1**

**5.** The diagrams below show villi in the small intestine of a mammal.

Diagram A
Section through the small intestine

Diagram B
A single villus

(a) State how the arrangement of villi, shown in **Diagram A**, increases the efficiency of absorption of digested foods.

_____ 1

(b) Name the two structures, labelled X and Y on **Diagram B**, which transport digested food away from the intestine.

X _____ 1

Y _____ 1

[Turn over

6. A volunteer was given 1 litre of water to drink. Every 30 minutes for the next three hours, urine was collected and its volume and salt concentration were measured.

The results are shown on the graph below.

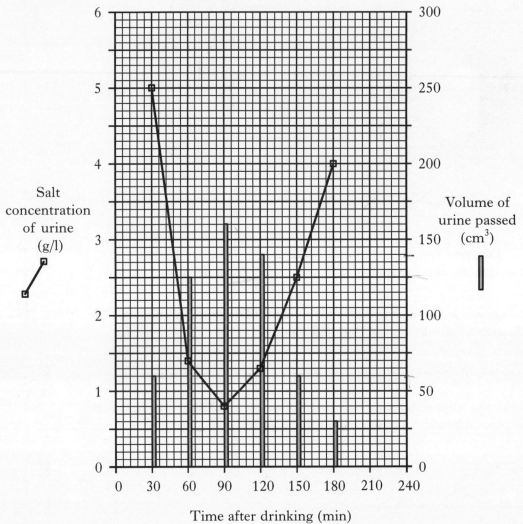

(a) What was the total volume of urine passed during this investigation?
*Space for calculation*

_____ cm³

(b) Using the data in the diagram, predict the salt concentration of a urine sample taken at 210 minutes.

Predicted salt concentration _____ g/l

6. (continued)

    (c) Between which two sample times would the volunteer's blood have contained the lowest concentration of ADH?

    *Tick the correct box.*

    0 – 30 minutes ☐

    30 – 60 minutes ☐

    60 – 90 minutes ☐

    90 – 120 minutes ☐

    (d) Describe the relationship between the volume of urine passed and its salt concentration.

    _____

    _____

7. Pike, roach and trout are freshwater fish which reproduce by external fertilisation of their eggs.

   Adult pike range from 100–150 cm in length and each female produces an average of 100 000 eggs which are each 2·5 mm in diameter.

   Adult trout and roach are each from 25–40 cm long. Roach produce the same number of eggs as pike on average, whilst trout produce only 1000 eggs per female.

   Trout produce the biggest eggs at 5 mm diameter, whilst roach eggs are only 1 mm.

   (a) Complete the following table by adding suitable column headings and data using the above information.

   | Fish | | | |
   |---|---|---|---|
   | Pike | | | |
   | Trout | | | |
   | Roach | | | |

   (b) Which species has the greatest chance of successful development?
   Give a reason for your answer.

   Species _____

   Reason _____
   _____

   (c) Fertilisation in land-living animals is internal.
   Explain the importance of this.

   _____
   _____

**8.** Read the passage below and answer the questions which follow it.

Birds migrate for the breeding season to areas with good food supplies and nesting places. They go elsewhere for the winter because conditions are no longer suitable. We see millions of insect-eaters such as swifts and swallows moving into Britain in summer, but migrating south to warmer climates in winter because they cannot survive without insect food. Resident species such as robins and blackbirds eat insects in summer and switch to a different food resource in winter and so do not migrate.

Some wildfowl and waders need to leave Britain in summer to breed but migrate here in winter to feed on the invertebrates present in our estuaries. Other migrants, both in spring and autumn, use our islands as stopovers to feed during their long migrations north and south.

There have been changes in bird distribution relating to factors like climatic changes. This has probably been responsible for redwings and fieldfares, which are normally migrants, establishing resident populations in Britain.

(a) Give **one** reason why some birds migrate to Britain to breed.

_____

(b) Give **one** reason why some birds migrate to Britain for the winter.

_____

(c) Name one resident and one migrant species which eats insects.

Resident species _____

Migrant species _____

(d) Explain why some species may be seen in Britain for short periods at two different times of the year.

_____

(e) What explanation is suggested in the passage for the resident populations of redwings and fieldfares in Britain?

_____

[Turn over

**9.** The diagram below represents a cell in an early stage of mitosis.

Which of the following diagrams represents the chromosomes you would expect to find in the nuclei of the daughter cells at the end of mitosis?

*Tick the correct box.*

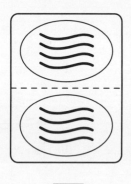 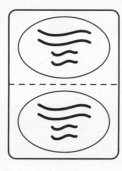 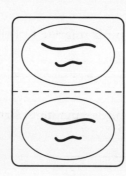 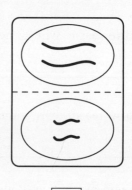

☐  ☐  ☐  ☐

**10.** The following experiment was set up.

Sodium hydroxide solution absorbs carbon dioxide from air.

Lime water turns from clear to cloudy in the presence of carbon dioxide.

Air is drawn through the apparatus from X to Y, passing through each flask in turn.

X air in → → Y air out

**Flask A** Sodium hydroxide solution  
**Flask B** Lime water  
**Flask C** Insect  
**Flask D** Lime water

(a) What should happen to the lime water in Flask B?

_____ 1

(b) (i) The lime water in Flask D turned cloudy after one hour.
Give a valid conclusion which could be drawn from this result.

_____

_____ 1

(ii) Predict how the results would differ if two insects were put into Flask C.

_____ 1

[Turn over

**11.** (a) (i) Tissue from an onion root was placed in water. The diagram below represents a cell from the tissue.

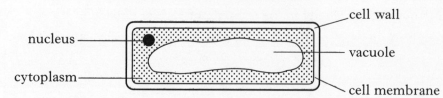

The tissue was then transferred into a very concentrated salt solution for one hour.

Complete the diagram below to show the appearance of the onion cell contents after this time.

(An additional diagram is available, if required, on page 27.)

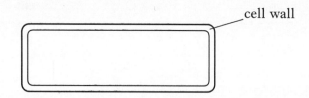

(ii) Underline **one** alternative in each group to make the following sentence correct.

In concentrated salt solution, water passes $\begin{Bmatrix} into \\ out\ of \end{Bmatrix}$ an onion cell from a region of $\begin{Bmatrix} high \\ low \end{Bmatrix}$ water concentration, to a region of $\begin{Bmatrix} high \\ low \end{Bmatrix}$ water concentration with the cell membrane acting as a $\begin{Bmatrix} selectively \\ fully \end{Bmatrix}$ permeable membrane.

(b) Explain the importance of diffusion for an onion root cell.

_____

_____

**11. (continued)**

(c) Five cylinders of potato tissue were weighed and each was placed into a salt solution of a different concentration.

The cylinders were reweighed after one hour and the results are shown in the following table.

| Salt solution | A | B | C | D | E |
|---|---|---|---|---|---|
| Initial mass of potato cylinder (g) | 10 | 10 | 10 | 10 | 10 |
| Final mass of potato cylinder (g) | 12·6 | 11·2 | 10·1 | 9·4 | 7·0 |

(i) The potato cylinders were blotted dry before each weighing. Suggest a reason for this.

_____

_____

(ii) Which salt solution had the highest water concentration?

Salt solution _____

(iii) Calculate the percentage decrease in mass of the potato cylinder in salt solution D.

*Space for calculation*

_____ %

**12.** (a) The grid below is about breathing and lungs.

| A trachea | B mucus | C diaphragm | D cilia |
|---|---|---|---|
| E air sacs | F bronchi | G rib cage | H capillaries |

Use letters from the boxes to complete the following.

(i) Identify two structures which are supported by rings of cartilage.

Letter ____ and letter ____

(ii) Identify two structures which are used to change the volume of the lungs during breathing.

Letter ____ and letter ____

(iii) Identify two features which can help prevent dust from reaching the air sacs.

Letter ____ and letter ____

(b) The following graph shows the effect of a training programme on the number of blood capillaries in the heart muscle of an athlete.

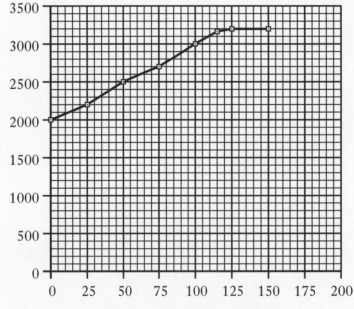

**12.** *(b)* (continued)

(i) Describe the relationship between the distance run per week and the number of capillaries in the heart muscle.

_____

_____   2

(ii) What was the percentage increase in the number of capillaries per $mm^3$ of heart muscle when the distance run each week was increased from 50 to 100 km?

*Space for calculation*

_____ %   1

*(c)* (i) Training increases the efficiency of the heart.

Explain how an increased number of capillaries in the heart muscle contributes to its efficiency.

_____

_____   1

(ii) In addition to improving the blood circulation, state **one** other way in which training improves the efficiency of the body.

_____   1

[Turn over

**13.** The diagram shows a ball and socket joint.

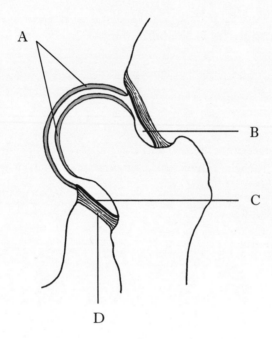

Complete the table with the letters, names and functions of the labelled structures in the joint.

| Letter | Name of structure | Function |
|---|---|---|
|  | synovial fluid |  |
| C |  | produces synovial fluid |
| A |  | cushions the joint |
|  |  | holds bones together |

3

**14.** An investigation was carried out to test the hypothesis that using both eyes increases the ability to judge distances.

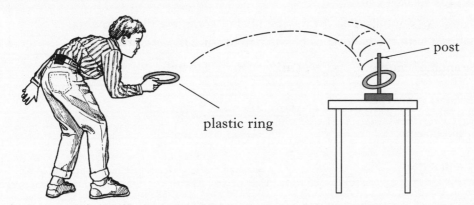

Four volunteers threw plastic rings at a post.

Each volunteer had 20 throws with no eyes covered, with one eye covered and with both eyes covered.

The results are shown on the table below.

| Volunteer | Number of successes out of 20 throws | | | |
|---|---|---|---|---|
| | no eyes covered | right eye covered | left eye covered | both eyes covered |
| 1 | 8 | 3 | 3 | 1 |
| 2 | 12 | 4 | 3 | 2 |
| 3 | 6 | 2 | 3 | 0 |
| 4 | 8 | 5 | 4 | 0 |
| Average | 8·50 | 3·50 | 3·25 | |

(a) Complete the table to show the average result with both eyes covered.
*Space for calculation*

(b) Name two variables concerning the apparatus for the experiment which must be kept the same throughout the investigation.

1 _____

2 _____

(c) Underline **one** alternative in each group to make the following statements correct.

The variable tested in the investigation was the

$\begin{Bmatrix} \text{diameter of the hoops} \\ \text{number of successful throws} \\ \text{number of eyes used} \end{Bmatrix}$. The hypothesis should be $\begin{Bmatrix} \text{accepted} \\ \text{rejected} \\ \text{modified} \end{Bmatrix}$.

DO NOT WRITE IN THIS MARGIN

Marks | KU | PS

**15.** (*a*) In an investigation into the inheritance of height in pea plants, true-breeding tall plants were crossed with true-breeding dwarf plants.

All the $F_1$ plants were tall.

(i) Using the symbols **T** and **t** for the alleles, complete the following diagram with the genotypes of the parents and the offspring.

Parental phenotypes **Tall** × **Dwarf**

Parental genotypes ____ ____

$F_1$ phenotype **Tall**

$F_1$ genotype ____ **1**

(ii) If a second generation of pea plants was produced by allowing the $F_1$ generation to self-cross, what would be the expected ratio of phenotypes?

*Space for working*

Expected $F_2$ ratio **Tall** : **Dwarf**

____ : ____ **1**

(iii) When the $F_2$ plants were counted, there were 720 tall plants and 180 dwarf plants.

Calculate the actual ratio of tall plants to dwarf plants.

*Space for calculation*

Actual $F_2$ ratio **Tall** : **Dwarf**

____ : ____ **1**

(iv) Explain why these results differ from the expected ratio.

________________________________________

________________________________________ **1**

**15. (continued)**

(b) Tallness and dwarfness in pea plants is an example of *discontinuous variation*. Explain the meaning of this term.

_____

_____

**1**

[Turn over

**16.** (a) The following list describes some of the stages in the production of human insulin by genetically engineered bacteria.

Stage 1 _____

Stage 2 Insertion of the insulin gene into the chromosomal material of suitable bacteria.

Stage 3 Bacteria reproduce rapidly, passing on the insulin gene.

Stage 4 _____

Stage 5 Extraction and purification of the insulin.

(i) In the spaces provided, describe stages 1 and 4.

(ii) Explain why there is an ever increasing need for insulin produced by bacteria.

_____

(b) Compared to selective breeding, state **one** advantage of genetic engineering as a way of improving the characteristics of a species.

_____

**17.** The graph shows the population growth of yeast cells in a fermenter.

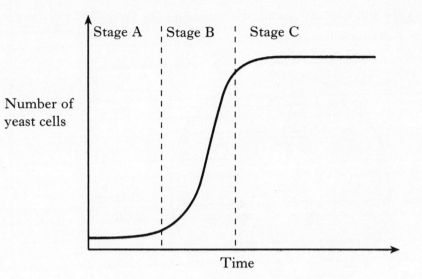

(a) Which stage on the graph shows the fastest population growth?

Stage _____

(b) Describe the changes in population growth shown in Stage C on the graph, and give a reason for the changes.

Changes _____

_____

Reason _____

(c) The fermenter was cleaned by steam sterilisation at 121 °C before it was used.

Name the structures, produced by bacteria and fungi, which could have survived if boiling water had been used for cleaning.

_____

[Turn over

**18.** A suspension of bacteria was spread evenly over the surface of a nutrient agar in a petri dish.

A multidisc containing six different antibiotics was placed on the agar. The diagram below shows the appearance of the petri dish after it had been incubated for two days.

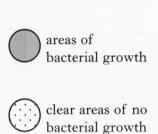
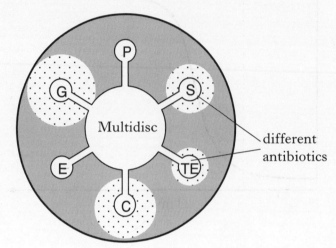

(a) Complete the table below to record the effectiveness of each antibiotic.

| Antibiotics which had some effect | Antibiotics which had no effect |
|---|---|
|  |  |

1

**18. (continued)**

(b) The table below shows the results from a similar investigation with a different bacterium.

| Antibiotic | Diameter of clear area (mm) |
|---|---|
| P | 0 |
| S | 4·1 |
| TE | 2·2 |
| C | 5·0 |
| G | 4·3 |
| E | 0·5 |

(i) Use the information from the table to complete the Y-axis and plotting of the bar chart on the grid below.

(An additional grid is available, if required, on page 28.)

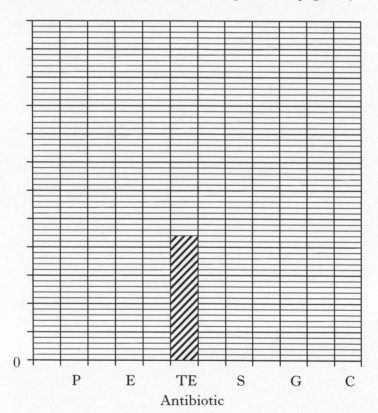

(ii) Suggest the most effective antibiotic to use in the treatment of a patient infected with this bacterium.

Antibiotic _____

(c) Explain why a range of antibiotics is needed in the treatment of bacterial diseases.

_____

_____

**19.** (a) Commercial brewers provide the best growing conditions for yeast.

Draw clear lines to link the growing condition required by yeast with the method used to provide it.

| Growing condition | Method |
|---|---|
| Food supply | sterilisation |
| Suitable temperature | thermostats |
| Lack of competition | germinating barley grains |

(b) Yeast cells can be measured in micrometres.
1 millimetre (mm) = 1000 micrometres (μm).
If 20 yeast cells together measure 1 mm, what is the average size **in micrometres** of **one** yeast cell?

*Space for calculation*

_____ μm

[END OF QUESTION PAPER]

ADDITIONAL GRAPH PAPER FOR QUESTION 3(b)

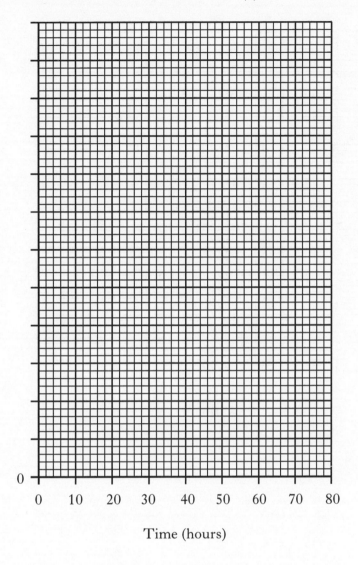

Time (hours)

ADDITIONAL DIAGRAM FOR QUESTION 11(a)(i)

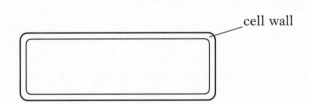

[Turn over

ADDITIONAL GRID FOR QUESTION 18(b)(i)

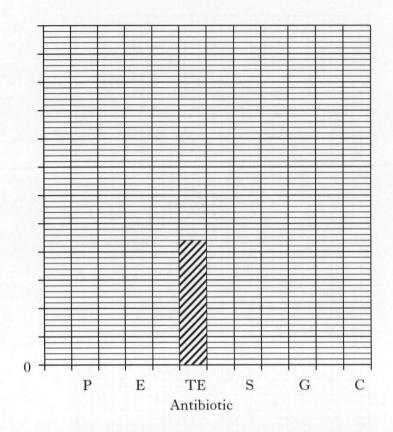

**2002** | Credit

Official SQA Past Papers: Credit Biology 2002

**FOR OFFICIAL USE**

C

KU | PS

**Total Marks**

**0300/402**

NATIONAL QUALIFICATIONS 2002

FRIDAY, 24 MAY 10.50 AM – 12.20 PM

**BIOLOGY STANDARD GRADE**
Credit Level

**Fill in these boxes and read what is printed below.**

Full name of centre

Town

Forename(s)

Surname

Date of birth
Day Month Year

Scottish candidate number

Number of seat

1 All questions should be attempted.

2 The questions may be answered in any order but all answers are to be written in the spaces provided in this answer book, and must be written clearly and legibly in ink.

3 Rough work, if any should be necessary, as well as the fair copy, is to be written in this book. Additional spaces for answers and for rough work will be found at the end of the book. Rough work should be scored through when the fair copy has been written.

4 Before leaving the examination room you must give this book to the invigilator. If you do not, you may lose all the marks for this paper.

1. The diagram below shows part of a food web in the Irish Sea.

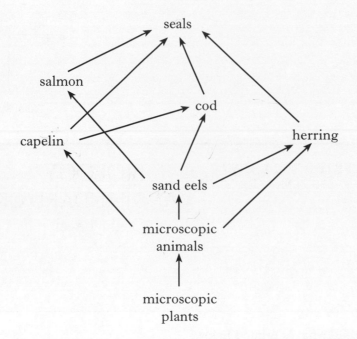

(a) Two food chains from the food web are made up of four populations of organisms.

Complete both of these food chains in the spaces below.

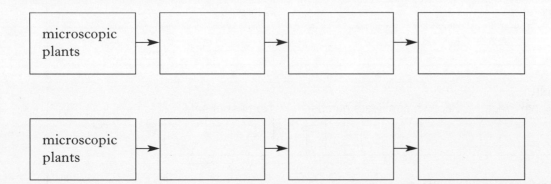

**1. (continued)**

(b) Predict the effect on the population of (i) capelin, and (ii) seals if the sand eels were removed from this food web.

(i) <u>Underline</u> your prediction and explain your choice.

Capelin would { increase / decrease / stay the same }.

Explanation _____

_____

(ii) <u>Underline</u> your prediction and explain your choice.

Seals would { increase / decrease / stay the same }.

Explanation _____

_____

(c) (i) What term is used to describe a diagram that shows the total mass of organisms present at successive levels of a food chain?

_____

(ii) Which of the following may **not** decrease at each successive level of some food chains?

<u>Underline</u> the correct answer.

energy      numbers      biomass

[Turn over

**2.** (a) Tropical rain forests are estimated to contain more than half of the Earth's existing species of plants and animals, many of which have not yet been studied.

Rain forests are being destroyed, leading to a reduction in the number of species. This has possible consequences for humans and other animals.

Describe **one** such possible consequence for humans.

_____

_____

(b) The diagrams show two types of structures found in plants.

A          B

(i) Which structure would be found in the phloem?

_____

(ii) Xylem helps to support a plant. State **one** other function of xylem.

_____

**3.** The graph below shows the average dry mass of potato tubers and the shoots they produced over a period of eight weeks.

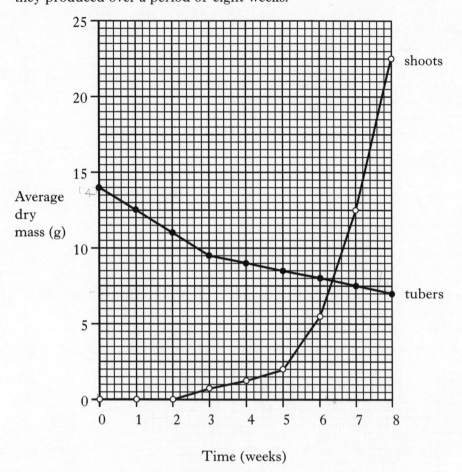

*(a)* The dry mass of the tubers decreased during the eight weeks.

Calculate the decrease as a percentage of their original dry mass.

*Space for calculation*

_____ %

*(b)* Why was the dry mass, rather than the fresh mass, of the tubers and shoots measured?

_____

_____

*(c)* Predict when the potato plants begin to photosynthesise and explain your answer.

Time from planting _____ weeks.

Explanation _____

_____

**4.** Dwarf bean plants were grown in pots of sand containing different masses of nitrogen fertiliser. Five pots were set up for each mass of fertiliser. After 10 weeks, the plants were dug up and their root nodules were removed, washed and weighed. The results are shown in the table.

| Mass of nitrogen fertiliser (g) | Average mass of root nodules per plant (g) |
|---|---|
| 0 | 5·3 |
| 0·2 | 1·6 |
| 1·0 | 0·8 |
| 5·0 | 0·4 |
| 10·0 | 0·1 |

(a) (i) On the grid below, complete a **line graph** to show the effects of increasing the mass of nitrogen fertiliser on the mass of root nodules formed by the bean plants.
(An additional grid, if required, will be found on page 27.)

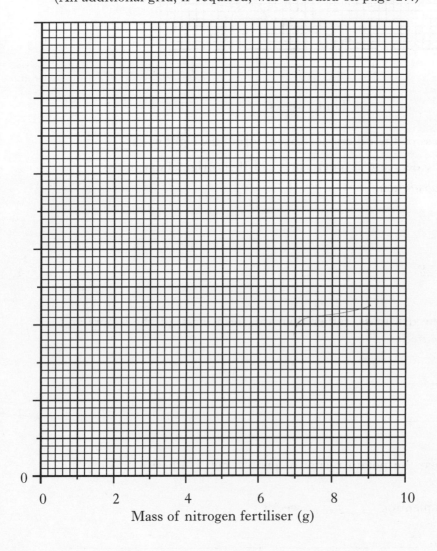

4. (a) (continued)

(ii) What effect does increasing the mass of nitrogen fertiliser have on the mass of root nodules formed per plant?

_____ 1

(b) Why was it good experimental technique to set up five pots for each mass of fertiliser?

_____ 1

(c) What type of bacteria is found in the root nodules of the dwarf bean plants?

_____ 1

[Turn over

**5.** (a) The experiment below was set up to investigate the effect of copper on the growth of one species of grass plant.

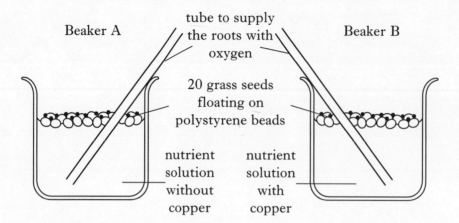

The length of the roots was measured every five days. The results are shown in the table.

| Day | Average length of roots (mm) | |
|---|---|---|
| | Beaker A | Beaker B |
| 0 | 0 | 0 |
| 5 | 13 | 9 |
| 10 | 15 | 13 |
| 15 | 19 | 13 |
| 20 | 22 | 14 |
| 25 | 30 | 18 |

(i) Calculate the average increase in root length per day, during the 25 days, for the grass plants in Beaker A.

*Space for calculation*

Average increase in root length _____ mm per day.

(ii) Calculate the simplest whole number ratio of average length on day 25 for the roots of the plants in Beaker A to those in Beaker B.

*Space for calculation*

_____ : _____
Beaker A    Beaker B

**5.** **(a)** **(continued)**

(iii) Describe the difference which copper makes to the growth of the grass plants.

_____

(iv) Beaker A is a control. What is the purpose of the control in this experiment?

_____

_____

(b) A similar experiment was carried out to investigate the effect of copper on the growth of a different species of grass plant.

State **two** precautions that would have to be taken to ensure that a valid comparison could be made between the two experiments.

1 _____

2 _____

[Turn over

**6.** (a) The diagram below represents a kidney nephron.

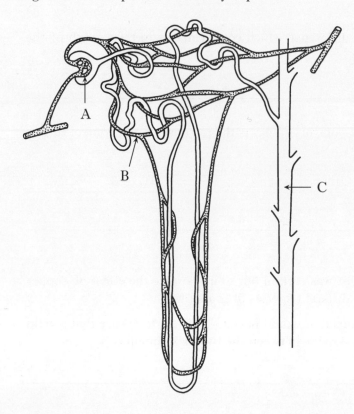

Complete the table by adding the correct letters, name and function.

| Letter | Name | Function |
|--------|------|----------|
|  | Collecting duct | Collects urine |
| A |  | Filters the blood |
|  | Blood capillary |  |

6. (continued)

(b) Dialysis is a process by which waste products are removed from the blood. The following information refers to artificial dialysis used as treatment for kidney failure.

**Haemodialysis**. Blood is removed from a vein in the forearm and passed into a "kidney machine". A synthetic membrane separates the blood from dialysis fluid into which impurities from the blood diffuse. This treatment lasts for five hours and is required three times per week.

**Peritoneal Dialysis**. The natural membrane (called the peritoneum) lining the abdomen is used to filter waste from the blood vessels that surround the peritoneum. Three times each day, fluid is run through a plastic tube into the abdomen and left for four hours. The fluid is then drained out and fresh fluid is run in to continue the process.

(i) Complete the table to summarise this information.

| Name of treatment | Type of membrane (natural or synthetic) | How often the treatment is required |
|---|---|---|
|  |  |  |
|  |  |  |

(ii) One of the impurities removed from the blood is urea.
From which food component is urea produced?

_____

(iii) People with kidney failure can be given a kidney transplant. Give **one** benefit to the patient of having a kidney transplant compared to dialysis.

_____

[Turn over

**7.** The diagram shows apparatus used to compare the energy content of dried food samples. A sample is weighed and then burned. The energy in the food is converted to heat and this is measured from the rise in temperature of the water.

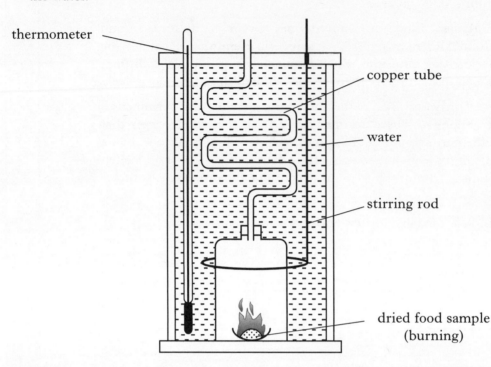

(a) (i) Why is it necessary to weigh the food samples before burning?

_____ 1

(ii) Explain the need for the following features of the apparatus.

1 A rod to stir the water _____

_____ 1

2 The copper tube is thin-walled and coiled _____

_____ 1

**7.** *(a)* **(continued)**

(iii) The following table shows results obtained using this apparatus.

Each food sample had a dry weight of 5 g.

| Food type | Initial temperature (°C) | Final temperature (°C) | Rise in temperature (°C) | Energy value (Joules/g) |
|---|---|---|---|---|
| olive oil | 23 | 45 | 22 | 35 |
| potato | 23 | 32 | 9 | 14 |
| lean meat | 25 | | | 14 |

Complete the table by inserting the final temperature and the rise in temperature for the sample of lean meat.

*Space for calculation*

(iv) This apparatus always gives an underestimate of the energy content of the foods tested.

Suggest a possible source of this error.

_____

*(b)* Which of the main types of food components, carbohydrates, fats or proteins, contains the most chemical energy per gram?

_____

*(c)* Name the process by which the chemical energy of food is released in a cell.

_____

[Turn over

**8.** The drawing represents part of a root tip as seen under high magnification.

(a) (i) What name is given to the type of cell division that can be seen in some of the cells?

_____

(ii) Describe what is happening in cells Y and Z.

Cell Y _____

_____

Cell Z _____

_____

(iii) Daughter cells produced by this type of cell division contain the same number of chromosomes as their parent cell. Explain the importance of this.

_____

_____

8. **(continued)**

   (b) (i) The process of cell division is controlled by many specific enzymes. Explain the term *specific* as used in this context.

   _____

   _____

   (ii) Enzymes have an optimum temperature and pH. Explain the meaning of the word *optimum*.

   _____

   _____

   [Turn over

9. The diagram shows a human hip joint.

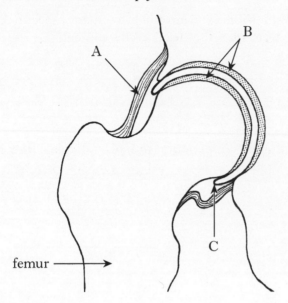

(a) Complete the table below by inserting the correct letters, name and functions.

| Letter | Name | Function |
|---|---|---|
| B |  | cushions the bone |
|  | synovial membrane |  |
|  | ligament |  |

(b) Decide if each of the following statements about the breathing system is **True** or **False** and tick (✓) the appropriate box.

If the statement is **False**, write the correct word in the **Correction** box to replace the word underlined.

| Statement | True | False | Correction |
|---|---|---|---|
| The trachea is supported by rings of <u>lignin</u>. |  |  |  |
| The air passages are lined by tiny hair-like <u>cilia</u>. |  |  |  |
| Special cells produce sticky <u>plasma</u> which prevents dust entering the lungs. |  |  |  |

9. (continued)

(c) The following statements refer to gas exchange between the blood capillaries and the air sacs in the lungs.

1 Carbon dioxide diffuses in.
2 Carbon dioxide diffuses out.
3 Oxygen diffuses in.
4 Oxygen diffuses out.

Complete the tables by inserting the number of each statement in the correct box.

| Air sacs | |
|---|---|
| | |

| Blood capillaries | |
|---|---|
| | |

(d) The following grid contains terms which refer to parts of the nervous system.

| A | B | C | D |
|---|---|---|---|
| spinal cord | touch receptor | relay nerve cell | medulla |
| E | F | G | H |
| cerebrum | cerebellum | motor nerve cell | sensory nerve cell |

Use the letters from the grid to identify the following.

(i) Part of the brain concerned with balance　_____

(ii) Structure which carries information from the sense organs　_____

(iii) Structure which carries information across the spinal cord during a reflex action　_____

(iv) Part of the brain which controls breathing and heart rates　_____

[Turn over

**10.** Read the following passage and answer the questions which follow it.

**All The Better To See You With**, adapted from J. Marsden, *Biological Sciences Review*, Vol **8**, 1995

Aqueous humour is a clear fluid that fills the front of the eye. Light passes through it before reaching the lens. It is constantly being made and drained away and supplies the metabolic needs of the lens and the cornea which have no blood supply. Aqueous humour contains glucose, amino acids and dissolved gases. Its pressure supports the eyeball and helps the eyeball to keep its shape.

Glaucoma occurs when the pressure inside the eye rises above normal. If not controlled, the pressure can squeeze the blood vessels in the eye. The increased pressure is usually due to problems with the drainage of the aqueous humour rather than too much being made. The effect is that the optic nerve is damaged due to decreased blood flow and poor oxygen supply resulting in loss of vision.

Chronic glaucoma results from a small rise in pressure over a long period of time. Sufferers feel no pain but the optic nerve is slowly damaged and peripheral vision is gradually reduced. This type of glaucoma is often discovered during a routine eye test. Families of glaucoma sufferers are able to obtain free eye tests. Drugs, in the form of eye drops, are used to increase the drainage of aqueous humour.

Acute glaucoma is a massive and rapid increase in the internal pressure of the eyeball caused by the iris blocking the drainage mechanism of the aqueous humour. It causes severe pain and loss of vision. A laser beam is used to form a hole in the iris to make a new drainage channel. People tend to get one type of glaucoma or the other, but not both.

(a) Why must the fluid of the aqueous humour be clear?

_____ 1

(b) How will carbon dioxide, produced by the respiring tissue of the cornea, be removed from the cornea?

_____ 1

(c) What is the usual cause of increased pressure in the eyeball?

_____ 1

10. (continued)

(d) Explain how an increase in pressure inside the eye can cause damage to the optic nerve.

_____

(e) What information in the passage suggests that glaucoma has a genetic component?

_____

(f) Describe **one** difference between chronic glaucoma and acute glaucoma.

_____

[Turn over

11. The diagram below shows part of the carbon cycle.

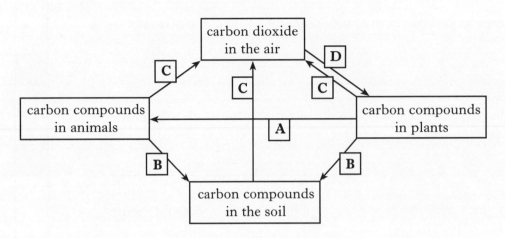

(a) Use **one** letter from the diagram to identify each of the stages in the table below.

| Stage | Letter |
|---|---|
| photosynthesis | |
| death and decay | |
| respiration | |

(b) Name a type of organism responsible for process B.

_____

(c) The following statements refer to the use of fossil fuels and nuclear fuels.

1   Contributes to acid rain.
2   Fuel supply likely to run out.
3   Waste material must be sealed in lead containers.
4   Releases carbon dioxide into the atmosphere.

Which statements refer to fossil fuels?

*Tick the correct box.*

1 and 2 only   ☐

2 and 3 only   ☐

1, 2 and 3     ☐

1, 2 and 4     ☐

**12.** In dogs, the difference between a coat which is the same colour all over, and a coat which has blotches of colour, is controlled by different forms of the same gene.

**B** (dominant) causes a blotched coat pattern and **b** (recessive) causes the same colour all over.

(a) What name is given to the different forms of the same gene?

_____

(b) A dog with the same coat colour all over mates with a blotched one. They have eight puppies, of which five have blotched coats and three are the same colour all over.

(i) What are the genotypes of the parent dogs?

Father _____ Mother _____

(ii) The predicted proportion of coat colours was equal numbers of each type. Explain why the actual numbers were different.

_____

_____

(c) Predict the genotypes and phenotypes of the puppies which would be produced if both parents had the same coat colour all over.

Genotype(s) _____

Phenotype(s) _____

(d) Is the variation in the dogs' coat pattern caused by the gene, continuous or discontinuous?

_____

**13.** Sugar can be produced from starch using an immobilised enzyme in the apparatus shown in the diagram below.

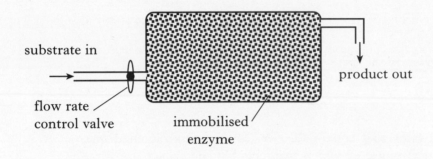

(a) (i) What is meant by the word "immobilised" in connection with enzymes?

_____

_____

(ii) Describe **one** advantage of using immobilised enzymes.

_____

_____

(b) Immobilised enzymes lose some of their activity over time. The graph shows the results of tests on the effectiveness of one immobilised enzyme produced by a Scottish company.

The tests were carried out at a temperature of 40 °C.

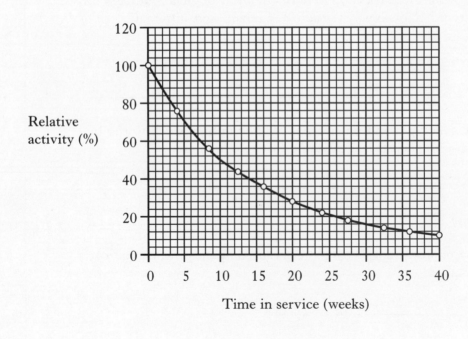

**13.** (*b*) (**continued**)

(i) What is the decrease in percentage relative activity of the enzyme after 26 weeks in service?

_____ %

(ii) When the enzyme is first used (time in service = 0 weeks), the substrate must remain in contact with it for six minutes to complete the reaction. How long would the enzyme require to be in contact with the substrate when the enzyme has been in use for ten weeks?

*Space for calculation*

_____ minutes

(iii) For the apparatus shown, how could the contact time between the enzyme and the substrate be increased?

_____

(iv) What would happen to the required contact time if the temperature was changed to 20 °C?

_____

(*c*) By using different enzymes in the same type of apparatus, it is possible to produce several synthetic antibiotics. Explain why a range of antibiotics is needed for the treatment of bacterial diseases.

_____

_____

[Turn over

**14.** The table gives information about the composition of some fatty foods.

| Food | Fat (g per 100g) | Cholesterol (mg per 100g) |
|---|---|---|
| Pork sausage | 25 | 60 |
| Cheddar cheese | 36 | 80 |
| Low fat spread | 82 | 0 |
| Butter | 84 | 225 |
| Milk | 4 | 15 |
| Egg | 12 | 450 |

(a) Express as a simple whole number ratio the mass of fat for milk, cheddar cheese and butter.

*Space for calculation*

_____ : _____ : _____
milk   cheddar   butter
       cheese

14. **(continued)**

(b) Complete the bar chart using information from the table.
(An additional grid, if needed, will be found on page 28.)

☐ Fat (g per 100 g)
▨ Cholesterol (mg per 100 g)

(c) What is the main difference in composition between low fat spread and butter?

_____

**15.** The table below contains information about some species of fish.

| Species | Size of scales | Number of dorsal fins | Barbels |
|---|---|---|---|
| Burbot | small | two | present |
| Pike | large | one | absent |
| Eel | small | one | absent |
| Grayling | large | two | absent |
| Miller's thumb | small | two | absent |

Use the information from the table to complete the boxes of the paired statement key below.

1  Large scales .................................................. go to 2
   Small scales .................................................. go to 3

2  One dorsal fin................................................ [Pike]
   Two dorsal fins.............................................. [Grayling]

3  Barbels present............................................ [Burbot]
   [Barbels absent] ........................... go to 4

4  [One dorsal fin] ........................... Eel
   [Two dorsal fins] ........................... Miller's thumb

[END OF QUESTION PAPER]

ADDITIONAL GRAPH PAPER FOR QUESTION 4(a)(i)

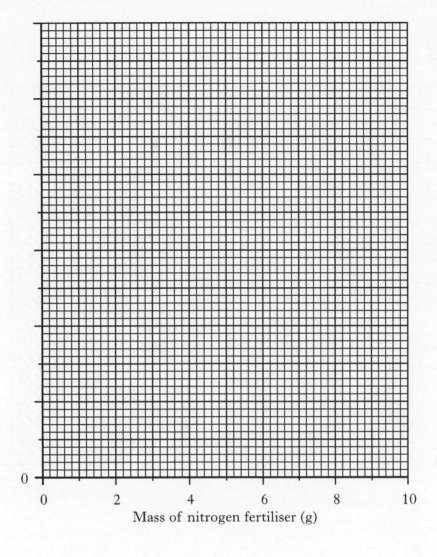

[Turn over

ADDITIONAL GRID FOR QUESTION 14(b)

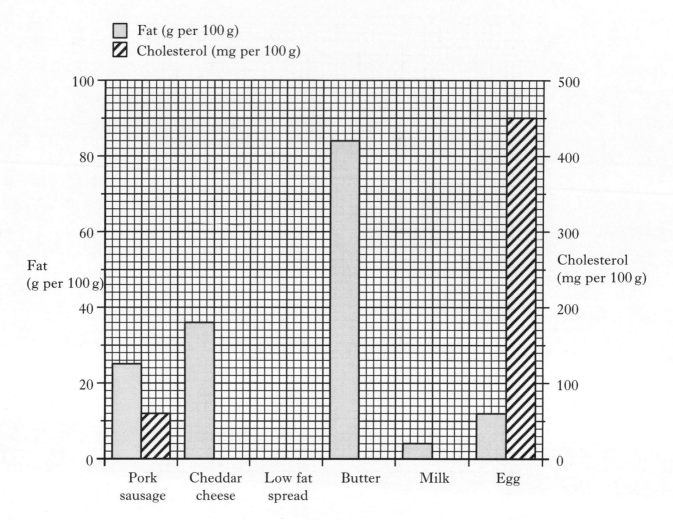

**2003** | Credit

**BLANK PAGE**

**FOR OFFICIAL USE**

KU   PS

Total Marks

## 0300/402

NATIONAL QUALIFICATIONS 2003

MONDAY, 26 MAY 10.50 AM – 12.20 PM

**BIOLOGY STANDARD GRADE** Credit Level

---

**Fill in these boxes and read what is printed below.**

Full name of centre

Town

Forename(s)

Surname

Date of birth
Day Month Year

Scottish candidate number

Number of seat

1 All questions should be attempted.

2 The questions may be answered in any order but all answers are to be written in the spaces provided in this answer book, and must be written clearly and legibly in ink.

3 Rough work, if any should be necessary, as well as the fair copy, is to be written in this book. Additional spaces for answers and for rough work will be found at the end of the book. Rough work should be scored through when the fair copy has been written.

4 Before leaving the examination room you must give this book to the invigilator. If you do not, you may lose all the marks for this paper.

SCOTTISH QUALIFICATIONS AUTHORITY

DO NOT WRITE IN THIS MARGIN

Marks KU PS

1. The diagram shows the results of a survey of seaweeds on a rocky Scottish shore. Starting at the highest tide level, square quadrats were placed every 5 metres in a line down the shore. Four species of seaweed were rated as absent, scarce or abundant in each quadrat.

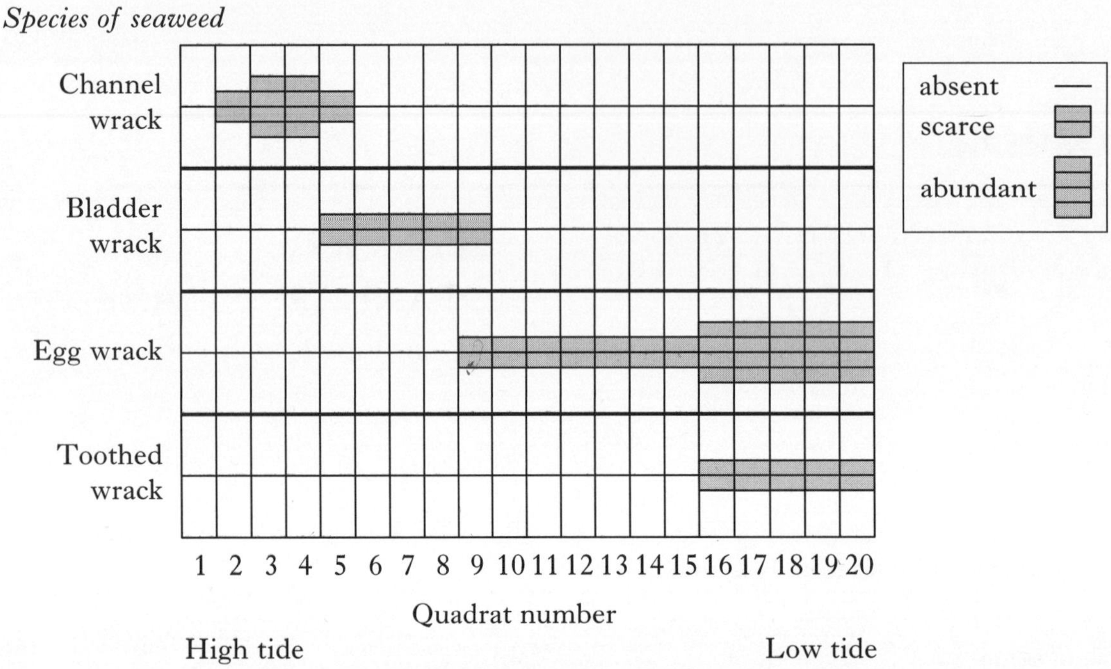


(a) (i) How many species of seaweed were found in quadrat number 9?

____________________ **1**

(ii) How many of the quadrats contained more than one species of seaweed?

____________________ **1**

(iii) Which species of seaweed spends least time covered by water?

____________________ **1**

(iv) What percentage of all the quadrats included egg wrack?

*Space for calculation*

__________ % **1**

1. (continued)

    (b) Suggest **one** abiotic factor that might affect the distribution of the seaweed species on the shore.

    _____ 1

    (c) Suggest **one** possible source of error in the sampling procedure and explain how it might be minimised.

    Source of error _____

    How to minimise it _____ 2

[Turn over

**2.** The diagram shows part of the human digestive system.

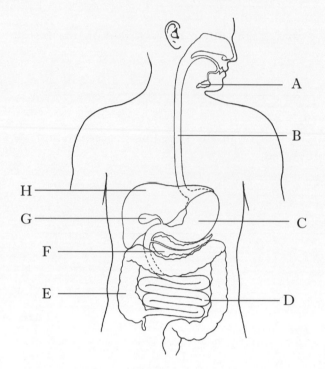

(a) Use letters from the diagram to answer the following.

   (i) Where is saliva produced? _____   1

   (ii) Where is bile produced? _____   1

(b) Name an enzyme responsible for:

   (i) the breakdown of fat. _____   1

   (ii) the breakdown of protein. _____   1

(c) Explain how contraction of muscles in the stomach wall speeds up digestion.

   _____   1

(d) The following statements refer to the process of peristalsis.

   *Tick the boxes of **two** correct statements.*

   Muscles in front of the food contract. ☐

   Muscles in front of the food relax. ☐

   Muscles behind the food contract. ☐

   Muscles behind the food relax. ☐   1

3. The table gives the partial composition of various types of milk.

| Type of milk | Mass of component per 100 cm³ | | | |
|---|---|---|---|---|
| | Protein (g) | Carbohydrate (g) | Fat (g) | Calcium (mg) |
| Skimmed | 3·4 | 5·0 | 0·1 | 124 |
| Semi-skimmed | 3·4 | 5·0 | 1·7 | 122 |
| Whole | 3·5 | 4·7 | 3·6 | 119 |

(a) (i) Use the information from the table to complete the bar chart below.
(An additional grid, if required, will be found on page 24.)

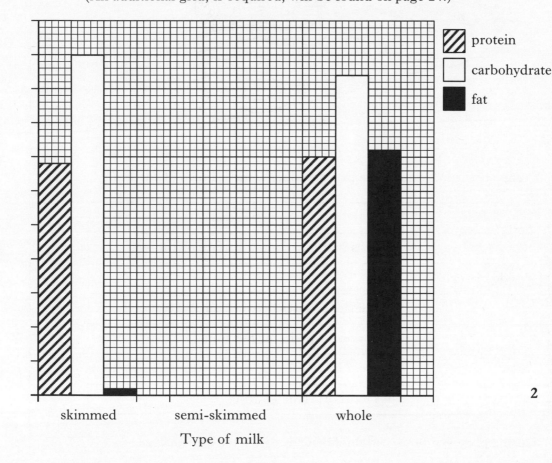

(ii) Which component shows the greatest variation in composition among the three types of milk?

_____

(b) The recommended daily intake of calcium is 800 mg. What percentage of this is supplied by 100 cm³ of skimmed milk?

*Space for calculation*

_____ %

**4.** The diagram shows a magnified view of the structure of a leaf.

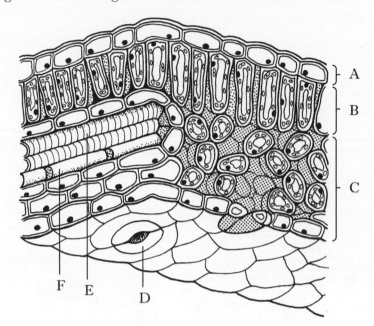

(a) Complete the following table that describes some features of the leaf.

| Letter | Name | Function |
|---|---|---|
| A | | Cells that form upper surface of the leaf |
| B | Palisade mesophyll | |
| C | | Exchanges gases between air and leaf cells |
| D | | Controls the size of the stoma |
| E | Xylem | |
| F | | Transports glucose from the leaf |

3

(b) During photosynthesis, carbon dioxide is converted into glucose.

(i) Name the structural carbohydrate, formed from glucose, that is the main component of cell walls.

_____

1

(ii) Give **one** use the plant makes of the glucose, other than the formation of structural materials.

_____

1

(c) Name the structural material that strengthens xylem vessels.

_____

1

4. (continued)

(d) The rate of photosynthesis can be limited by different factors.

Draw one line from each set of conditions to the factor that would be limiting photosynthesis.

*Set of conditions*

- High on a mountain on a sunny winter day
- The middle of a corn field on a warm bright day with no wind
- Late evening of a warm breezy summer day in a forestry plantation

*Factor limiting photosynthesis*

- Light intensity
- Wind speed
- Carbon dioxide availability
- Temperature

2

[Turn over

**5.** Potato cylinders of equal mass were placed in separate test tubes, as shown in the diagram.

salt solution
potato cylinder

The tubes contained salt solutions of 0·5%, 1·0%, 1·5%, 2·0% and 3·0% concentrations.

After two hours the change in mass of each cylinder was measured. The results are shown in the table.

| Tube | Change in mass (g) | Salt solution (%) |
|---|---|---|
| A | −0·6 | |
| B | −0·5 | |
| C | −0·2 | 1·5 |
| D | +0·1 | |
| E | +0·2 | |

(a) Complete the table by adding the correct concentration of the salt solution in each tube.

(b) Which tube contained a solution with a water concentration closest to that of the potato cell sap?

Tube _____

(c) The original mass of each potato cylinder was 5 g.
Calculate the percentage change in mass for the cylinder in tube D.
*Space for calculation*

_____ %

5. (continued)

(d) Underline **one** alternative in each bracket to explain the results for Tube C.

Water moved { into / out of } the potato by osmosis from a higher water concentration { inside / outside } the potato to a lower water concentration { inside / outside } the potato.

(e) Why would it be good experimental technique to blot the potato cylinders dry before each weighing?

_____

_____

(f) How could the results of the experiment be made more reliable?

_____

[Turn over

**6.** The following is a diagram of the human ear.

(a) The structures labelled **R** detect movements of the head.

(i) Give the name of these structures.

_____

(ii) Describe the arrangement of the structures labelled **R** and explain how this arrangement helps with their function.

Arrangement _____

_____

Explanation _____

_____

6. (continued)

(b) The following bar chart shows sound frequencies some animals can produce and hear.

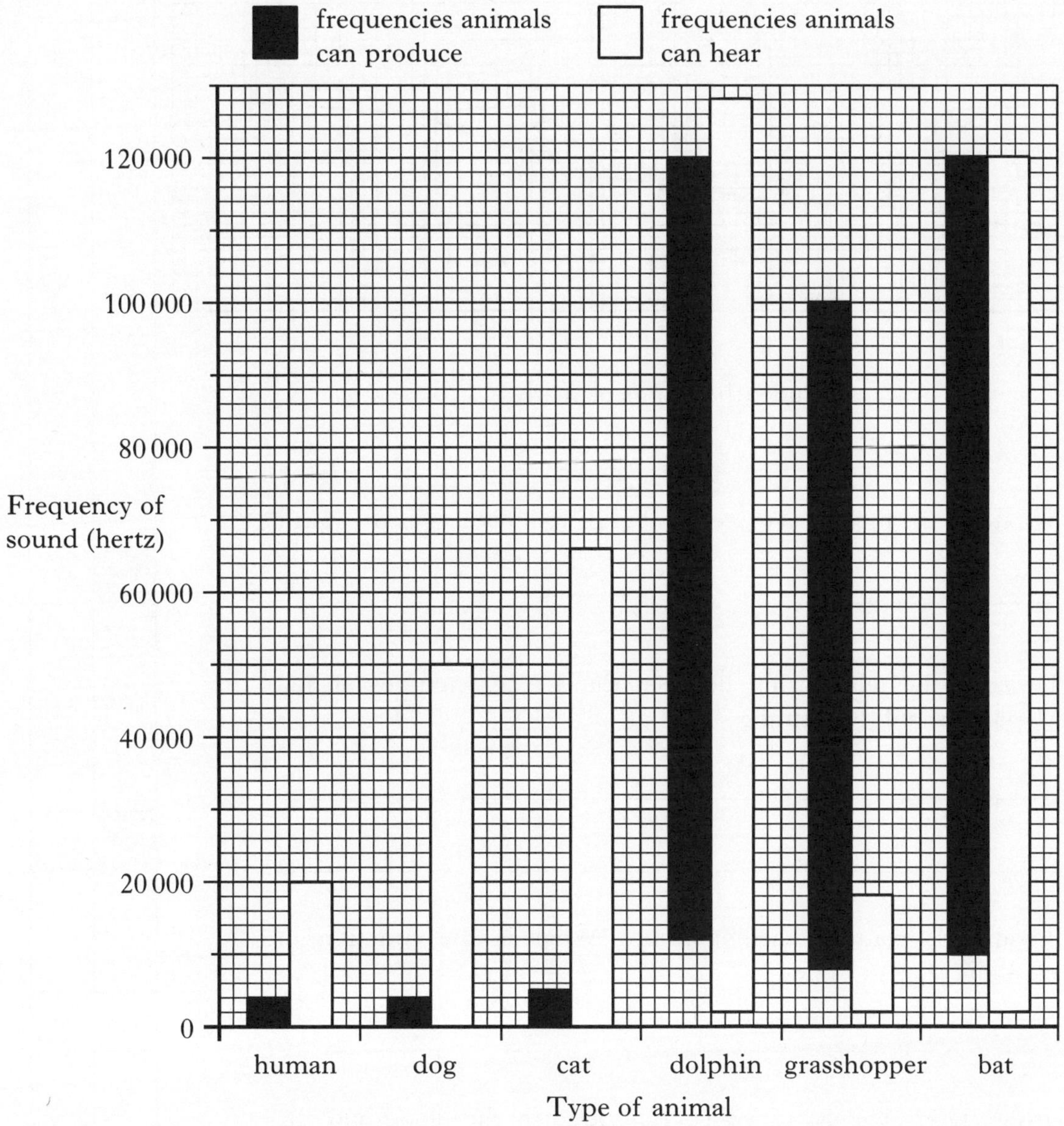

■ frequencies animals can produce    □ frequencies animals can hear

Frequency of sound (hertz) vs Type of animal (human, dog, cat, dolphin, grasshopper, bat)

Use information from the bar chart to answer the following questions.

(i) Which animal can hear the greatest range of sound frequencies?

_____

(ii) What is the lowest frequency of sound that can be heard by a bat?

_____ hertz

(iii) Name **all** the animals that can produce sounds which humans **cannot** hear.

_____

**7.** Groups of ten people who normally live at an altitude of zero metres were taken to stay at higher altitudes. The line graph shows the average number of red blood cells in each group after 100 days.

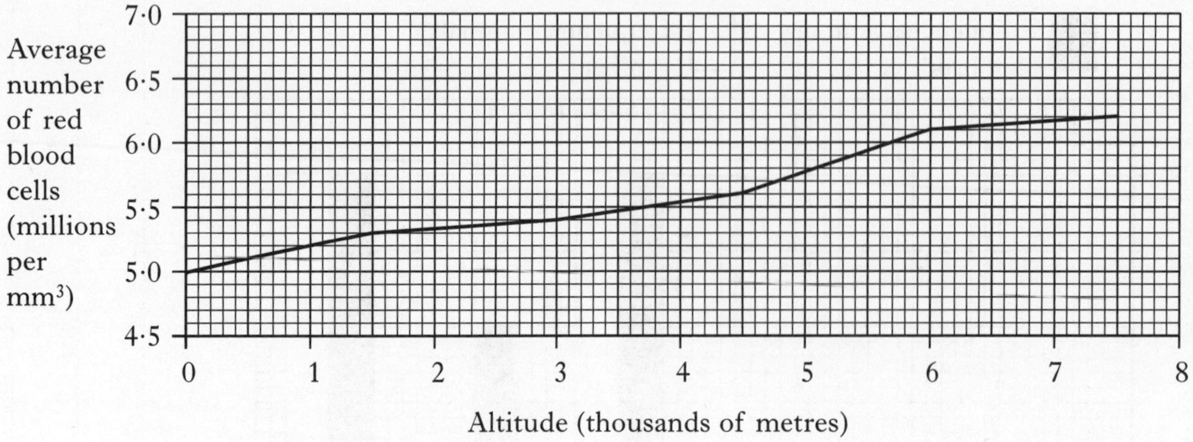

Average number of red blood cells (millions per mm³) vs Altitude (thousands of metres)

(a) (i) What was the highest average number of red blood cells per mm³?

_____

(ii) Describe the relationship between the average number of red blood cells and the altitude.

_____

_____

(b) Red blood cells contain haemoglobin. What is the function of haemoglobin?

_____

(c) Capillaries allow exchange of substances between the blood and the body tissues. Give **two** features of capillary networks that make this exchange efficient.

1 _____

2 _____

8. The graph shows the effect of temperature on the enzyme catalase.

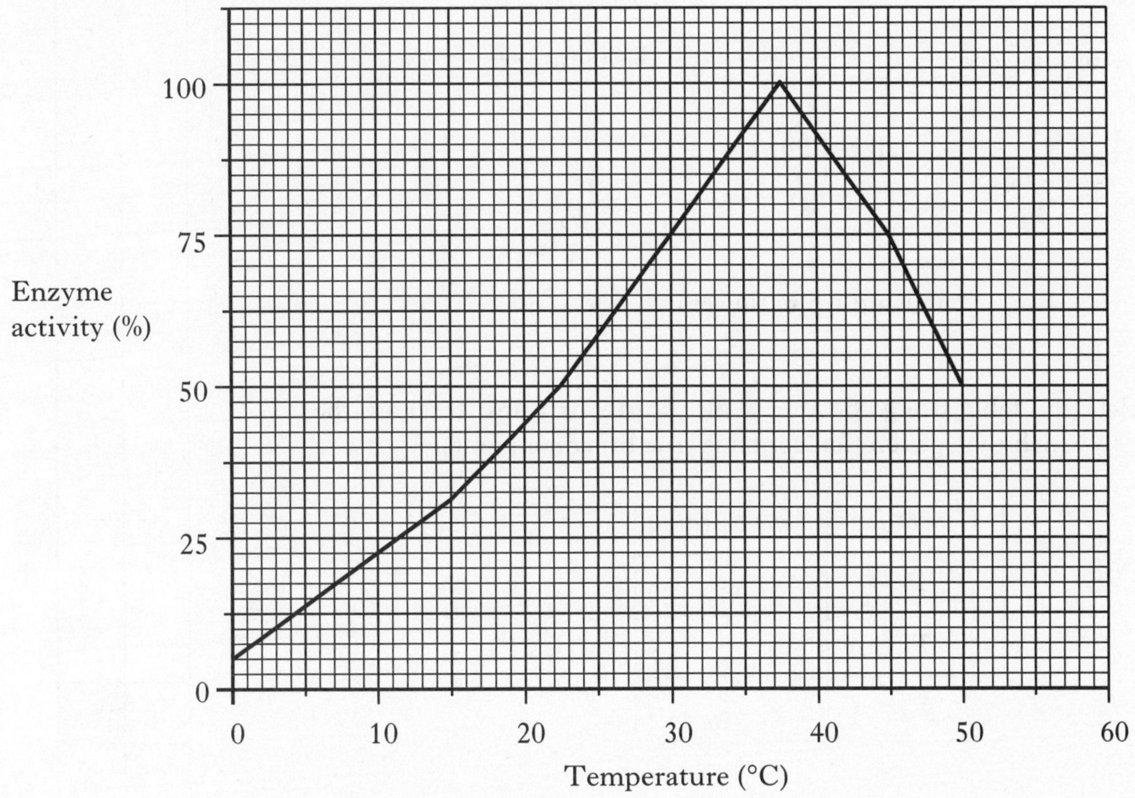

Enzyme activity (%) vs Temperature (°C)

(a) Between which **two** temperatures was there the greatest overall increase in enzyme activity?

*Tick the correct box.*

0 °C to 10 °C ☐

10 °C to 20 °C ☐

20 °C to 30 °C ☐

30 °C to 40 °C ☐

(b) At which **two** temperatures was the enzyme activity 75% of its maximum?

_____ °C and _____ °C

(c) From the graph, predict the temperature at which the enzyme activity will reach zero.

_____ °C

(d) Catalase will only work on one substrate.

What word is used to describe this feature of an enzyme?

_____

9. The following cross was carried out using plants with either red or white coloured flowers.

P      red flowers      ×      white flowers

$F_1$      all red flowers

$F_1$      ×      $F_1$

$F_2$      red and white flowers

(a) Complete the table using the words **all**, **some** or **none** to show the extent to which each generation contains true breeding plants.

| Generation | True breeding |
|---|---|
| P | |
| $F_1$ | |
| $F_2$ | |

(b) (i) Predict the expected number of white flowered $F_2$ plants that would have been produced if 1488 red flowered plants were produced.

*Space for calculation*

_____ white flowered plants

(ii) The actual number of white flowered $F_2$ plants was different from the expected. Suggest a reason why this happened.

_____

_____

(c) Using the letters **R** for red flowers and **r** for white flowers, give the genotypes of the plants in the table below.

| Plant | Genotype |
|---|---|
| red flowered $F_1$ | |
| white flowered $F_2$ | |

DO NOT WRITE IN THIS MARGIN

Marks | KU | PS

9. **(continued)**

(*d*) **R** and **r** represent two forms of the same gene.

What are the two forms of a gene called?

____________________________ **1**

(*e*) Flower colour is an example of discontinuous variation.

What is meant by the term "discontinuous variation"?

________________________________________

________________________________________ **1**

**[Turn over**

**10.** Read the following passage and answer the questions that follow.

**Great Oaks from Little Acorns Grow**, adapted from the Royal Horticultural Society's *Encyclopaedia of Practical Gardening*

Seeds are a resting and survival stage in a plant's life cycle. A seed consists of an embryo, a food supply and a seedcoat.

There are a number of distinctions that can be made among seeds. Seeds vary in size from small, dust-like seeds, such as those from rhododendrons and lobelia, to large seeds, such as acorns, chestnuts and hazelnuts. The enormous variation in the size of seeds influences the success of their growth. Large seeds are produced in small numbers, germinate satisfactorily and establish well. Dust-like seeds have lower survival rates.

Seeds also vary in the materials used as a food store. Seeds that store food as carbohydrates, such as elderberries, marigolds and laburnum, are generally stable, long-lived and will withstand drying. Seeds that store food as fats or oils, for example peony, magnolia and chestnut seeds, do not survive storage or drying very well.

Survival of drying, however, is not just affected by the stored food. It also reflects the condition of the seedcoat and its ability to protect the seed. Plants, such as willows, with very poorly developed seedcoats survive for only very short periods, while those plants, such as sweet peas, laburnum and lupin, with very hard, impermeable seedcoats usually survive for considerable periods in a wide variety of conditions. Seeds of the Indian lotus have germinated after 1000 years in a peat bog.

(a) How would the number of seeds produced by rhododendrons, and their survival rate, compare with those of chestnuts?

Number of seeds _____

Survival rate _____

(b) Give all the information contained in the passage about the seeds of laburnum.

_____

_____

(c) Which would contain more stored energy per gramme of its food store, marigold or magnolia? Give a reason for your answer.

Seed _____

Reason _____

_____

**10. (continued)**

(d) What factor, other than the nature of the food store, does the passage mention as important in allowing seeds to survive dry conditions?

_____

(e) What may be deduced about the seedcoats of the Indian lotus?

_____

(f) Use the information **in the passage** to complete the table below by entering a tick to describe the size and type of food store for each seed. The line for acorns has been completed.

| Seed | Size | | | Food store | | |
| --- | --- | --- | --- | --- | --- | --- |
| | large | small | no information | carbohydrate | fat | no information |
| Acorn | ✓ | | | | | ✓ |
| Chestnut | | | | | | |
| Elderberry | | | | | | |
| Lobelia | | | | | | |
| Peony | | | | | | |

[Turn over

**11.** The diagram and table describe part of a sewage treatment works.

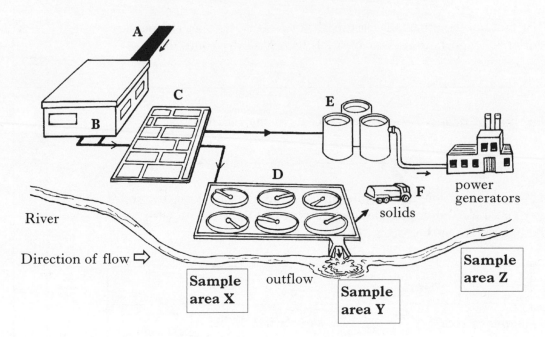

| Number | Process |
|---|---|
| 1 | Large pieces removed by mesh filter. |
| 2 | Solid material separated from liquid by allowing solids to settle. |
| 3 | Bacteria feed on solid organic material and produce biogas. |
| 4 | Wide range of micro-organisms feed on liquid waste material in aerobic conditions and decompose it to harmless products. |
| 5 | Waste materials from homes and factories. |
| 6 | Remaining solids dried and taken away to be used as fertiliser. |

(a) Complete the flow chart below by inserting the correct number from the table at each stage.

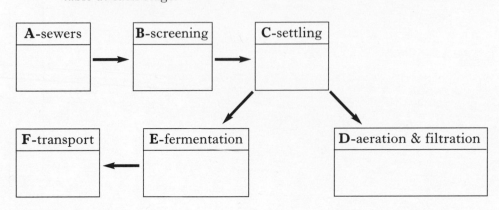

DO NOT WRITE IN THIS MARGIN

Marks | KU | PS

**11. (continued)**

(*b*) (i) Why is it important for aerobic conditions to be present during process 4?

______________________________________

______________________________________ **1**

(ii) Explain why a range of micro-organisms is needed to decompose sewage.

______________________________________

______________________________________ **1**

(*c*) The local authority checks for possible pollution caused from the sewage works by measuring the oxygen concentration of the river water and by monitoring indicator organisms.

(i) Which of the sample areas shown in the diagram would have the highest oxygen concentration if organic matter was present in the outflow?

*Tick the correct box.*

Sample area X ☐

Sample area Y ☐

Sample area Z ☐ **1**

(ii) Explain what is meant by an indicator species.

______________________________________

______________________________________ **1**

**[Turn over**

**12.** The apparatus shown below was used to investigate the effect of temperature on the rate of respiration in germinating peas.

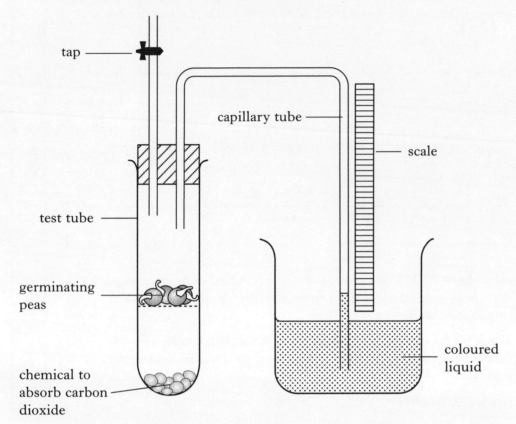

The test tube was placed in a water bath at 5 °C. The volume of oxygen used in respiration was measured by the movement of the coloured liquid in the capillary tube.

The experiment was repeated at different temperatures. The results are shown in the table.

| Temperature (°C) | Rate of respiration (cm³ oxygen used per hour) |
|---|---|
| 5 | 0·10 |
| 10 | 0·30 |
| 15 | 0·45 |
| 20 | 0·65 |
| 25 | 0·90 |
| 30 | 1·15 |
| 35 | 1·50 |
| 40 | 1·20 |
| 50 | 0·20 |

**12. (continued)**

(a) Draw a **line graph** of the results using an appropriate scale to fill most of the graph paper.
(Additional graph paper, if required, will be found on page 24.)

Temperature (°C)

(b) From the results, describe the relationship between temperature and the rate of respiration.

_____

_____

(c) A control experiment for this investigation used peas that had been boiled and then cooled.

(i) Explain the need for this control experiment.

_____

_____

(ii) Describe the expected results for the control experiment.

_____

_____

(d) If fresh plant leaves had been used instead of germinating peas, in the investigation, explain why the test tubes should be covered with black plastic.

_____

**13.** In cheese making both bacteria and fungi may be used.

(a) Underline **one** word in each bracket to explain what happens during the souring of milk for cheese making.

The pH of the milk {rises / falls} due to bacteria fermenting {lactose / glucose / maltose} sugar and producing {citric / lactic / nitric} acid.

(b) Blue cheese is made using a fungus that must be allowed to respire aerobically.

Other than carbon dioxide, which substance would be produced if the fungus respired anaerobically.

_____

(c) Temperature and pH are carefully controlled during cheese making to provide the optimum conditions for the enzymes involved.

Explain the meaning of the term "optimum conditions".

_____

_____

(d) The table gives information about five different cheeses.

| Type of cheese | Acid composition (%) |
|---|---|
| Cheddar | 0·60 – 0·70 |
| Cheshire | 0·60 – 0·70 |
| Leicester | 0·55 – 0·60 |
| Stilton | 1·10 – 1·30 |
| Wensleydale | 0·52 – 0·62 |

13. (d) (continued)

(i) Complete the chart below using information from the table.
(An additional grid, if required, will be found on page 25.)

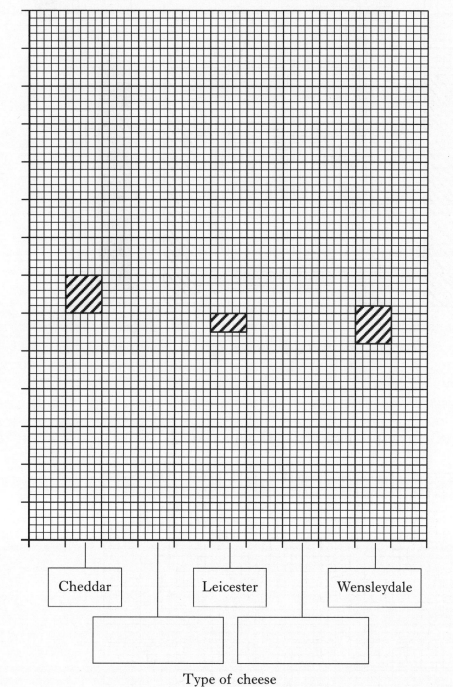

Type of cheese

(ii) Which cheese has the lowest pH?

_____

[END OF QUESTION PAPER]

ADDITIONAL GRID FOR QUESTION 3(a)(i)

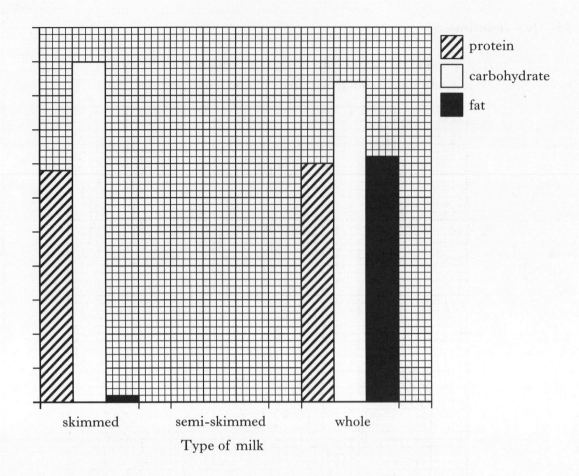

ADDITIONAL GRAPH PAPER FOR QUESTION 12(a)

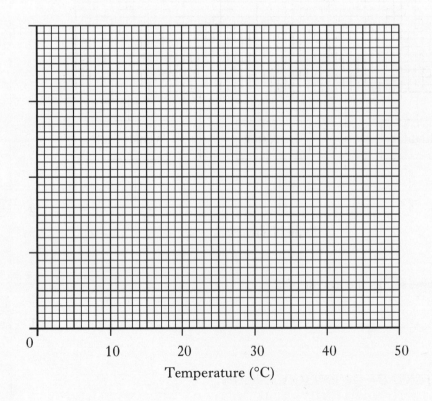

ADDITIONAL GRID FOR QUESTION 13(d)(i)

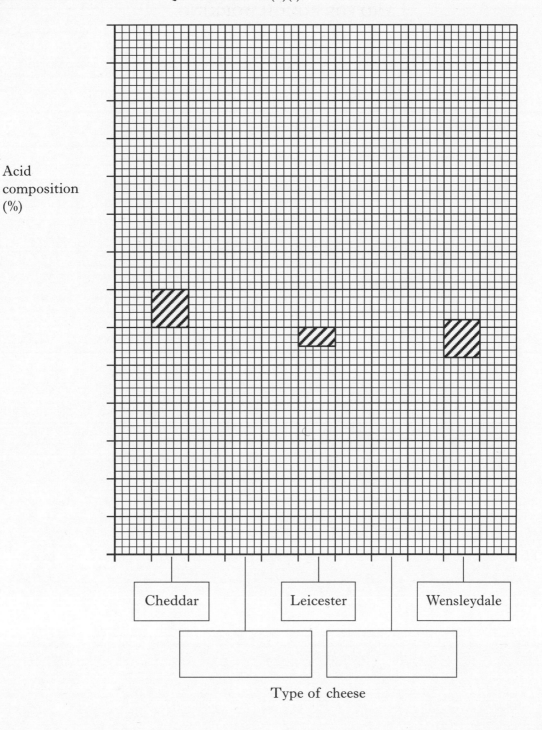

SPACE FOR ANSWERS
AND FOR ROUGH WORKING

**2004** | Credit

# FOR OFFICIAL USE

KU | PS

**Total Marks**

## 0300/402

NATIONAL QUALIFICATIONS 2004

WEDNESDAY, 19 MAY 10.50 AM – 12.20 PM

**BIOLOGY STANDARD GRADE** Credit Level

**Fill in these boxes and read what is printed below.**

Full name of centre

Town

Forename(s)

Surname

Date of birth
Day Month Year   Scottish candidate number   Number of seat

1 All questions should be attempted.

2 The questions may be answered in any order but all answers are to be written in the spaces provided in this answer book, and must be written clearly and legibly in ink.

3 Rough work, if any should be necessary, as well as the fair copy, is to be written in this book. Additional spaces for answers and for rough work will be found at the end of the book. Rough work should be scored through when the fair copy has been written.

4 Before leaving the examination room you must give this book to the invigilator. If you do not, you may lose all the marks for this paper.

**1.** The table contains information about five species of bat.

| Species | Wingspan (cm) | Roosting place | Flight |
|---|---|---|---|
| Pipistrelle bat | 19–25 | Trees and buildings | Fast and erratic |
| Leisler's bat | 25–33 | Trees and buildings | Fast and straight |
| Lesser Horseshoe bat | 19–25 | Buildings only | Fast and agile |
| Bechstein's bat | 25–33 | Trees and buildings | Slow and fluttering |
| Daubenton's bat | 19–25 | Trees and buildings | Fast and straight |

Use the information from the table to complete the boxes of the paired statement key below.

1   Wingspan 19–25 cm ......................... [          ]

    Wingspan 25–33 cm ......................... go to 3

2   Roosts in [          ] ............ Lesser Horseshoe bat

    Roosts in trees and buildings ........................... go to 4

3   Slow and fluttering flight ............................. [          ]

    Fast and straight flight .................................. [          ]

4   [          ] .......................... Pipistrelle bat

    [          ] .......................... Daubenton's bat

Marks: 3

**2.** The diagram shows part of a food web from the sea.

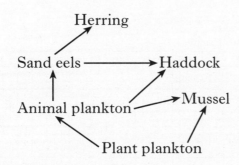

(a) Cod are fish that feed on young herring and sand eels.

Complete the food web to show the relationship of the cod to the other organisms in the food web. **1**

(b) Over fishing has led to a decrease in the numbers of haddock in the food web.

  (i) Explain why the population of animal plankton may **increase** if the haddock numbers are reduced.

  _____

  _____ **1**

  (ii) Explain why the population of animal plankton may **stay the same** if the haddock numbers are reduced.

  _____

  _____ **1**

(c) Underline **one** word in each pair to make the following sentences correct.

$\begin{Bmatrix} \text{Producers} \\ \text{Consumers} \end{Bmatrix}$ have the greatest biomass in a food chain. At each stage in the food chain, the biomass is $\begin{Bmatrix} \text{greater} \\ \text{smaller} \end{Bmatrix}$ than the stage before. The reason for this is that energy is $\begin{Bmatrix} \text{gained} \\ \text{lost} \end{Bmatrix}$ from each stage. **2**

[Turn over

3. The graph gives information about levels of sulphur dioxide in the air.

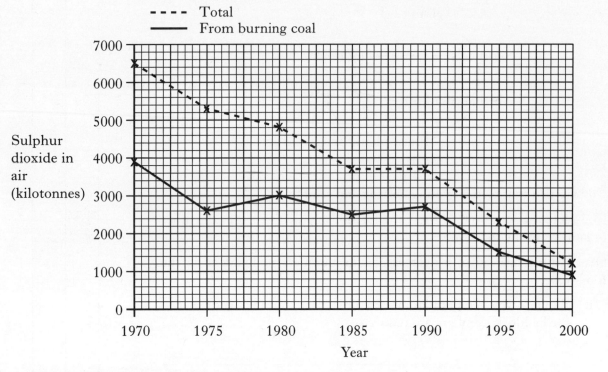

(a) The graph shows an overall reduction in the total level of sulphur dioxide in the air.

(i) During which 5 year period was there no reduction in the total level of sulphur dioxide in the air?

From _____ to _____

(ii) During which 5 year period was there the greatest decrease in the total level of sulphur dioxide in the air?

*Space for calculation*

From _____ to _____

(iii) In 1970, 3900 kilotonnes of sulphur dioxide were produced from burning coal. What percentage of the total is this?

*Space for calculation*

_____ %

**3.** *(a)* **(continued)**

(iv) Underline the correct option in the following sentence.

Between 1970 and 2000 the proportion of sulphur dioxide in the air which came from burning coal $\begin{Bmatrix} \text{increased} \\ \text{decreased} \\ \text{stayed the same} \end{Bmatrix}$.

*(b)* Sulphur dioxide levels in the air could be reduced by switching from coal-fired power stations to nuclear power stations.

Give one **disadvantage** of using nuclear power.

_____

_____

[Turn over

**4.** The table gives information about some bird species found on a deserted farm in Georgia, USA.
The bars show the presence of at least one breeding pair for each species.

| Bird species / Dominant plants | Time since the farm was deserted (years) | | | | | | | |
|---|---|---|---|---|---|---|---|---|
| | 1 | 3 | 15 | 20 | 25 | 60 | 100 | 150 |
| | Weeds | Grass | | Shrubs | | Pine trees | | Oak trees |
| Grasshopper sparrow | ■ | ■ | ■ | | | | | |
| Meadowlark | | ■ | ■ | ■ | | | | |
| Pine warbler | | | | ■ | ■ | ■ | ■ | |
| Cardinal | | | | ■ | ■ | ■ | ■ | |
| Wood thrush | | | | | | | | ■ |
| Field sparrow | | | | | | | | |

(a) Which bird species were found 15 years after the farm was abandoned?

_____ 1

(b) Which birds are likely to feed on seeds found in pine cones?

_____ 1

(c) If the farm was deserted in 1801, in what year did wood thrush first appear?

_____ 1

(d) Field sparrows were found on the farm 15 years after it was abandoned. There were no field sparrows recorded 45 years later. Add this information to the table by drawing in the correct bar. 1

**5.** (a) The following statements refer to the stages that occur after pollination.

A  Fertilisation takes place.
B  A pollen tube grows out from a pollen grain.
C  The ovule forms a seed and the ovary forms a fruit.
D  The pollen tube grows down through the stigma.
E  The male gamete moves towards the ovule.
F  The pollen tube grows through the ovary wall.

Use the letters of the statements to complete the sequence of stages.

B → [ D ] → [ F ] → E → [ A ] → [ C ]

(b) Plants can reproduce by sexual and asexual means.

Draw lines to link each method of reproduction with the advantages that each method provides.

*Method of reproduction*          *Advantage*

                                  Offspring obtains food and water from parent

   Sexual                         Seeds are dispersed

   Asexual                        Greater variation among offspring

                                  Pollination is not required

(c) The plants in a clone have been produced by asexual reproduction.

Give **one** other piece of information about the members of a clone.

_____

_____

[Turn over

**6.** (a) Complete the following table about the three major food groups.

| Type of food | Chemical elements present | Example of digestive enzyme | Product(s) of digestion |
|---|---|---|---|
| Carbohydrates | 1 carbon<br>2 hydrogen<br>3 oxygen | | |
| Fats | 1<br>2<br>3 | lipase | 1 fatty acids<br>2 glycerol |
| | 1 carbon<br>2 hydrogen<br>3 oxygen<br>4 nitrogen | | |

(b) The villi which line the small intestine each contain a lacteal and blood capillaries.

Give a brief description of the function of each of these structures.

Lacteal _____

_____

Blood capillaries _____

_____

**7.** The diagram shows a developing human fetus.

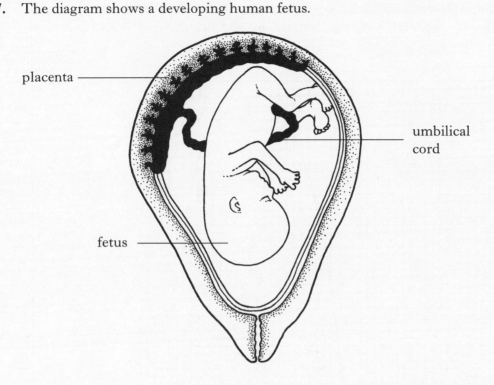

Complete the following table by putting a tick in the correct column to indicate the main direction of exchange for each of the following materials.

The first one has been done for you.

| Material | Direction of exchange | | |
|---|---|---|---|
| | Mother to fetus | Fetus to mother | No exchange |
| glucose | ✓ | | |
| amino acids | | | |
| blood | | | |
| oxygen | | | |
| urea | | | |
| carbon dioxide | | | |

2

[Turn over

**8.** The graph shows the relationship between daylength (hours of light in a 24 hour period) and the nest building activity of chaffinches and house sparrows.

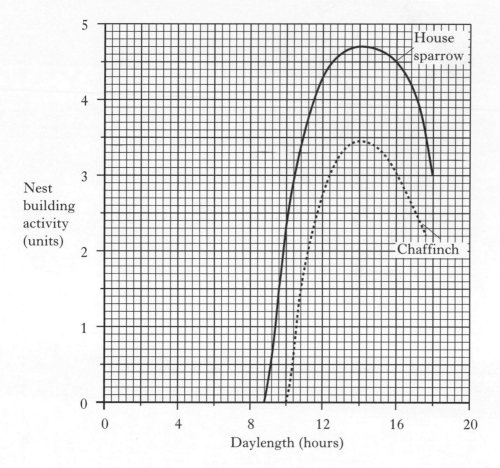

(a) Using data from the graph, describe the relationship between nest building activity and daylength for the chaffinch.

_____

_____

_____

(b) House sparrows begin nesting activity earlier than chaffinches. Suggest a benefit to the survival of house sparrows.

_____

_____

DO NOT WRITE IN THIS MARGIN

*Marks* | KU | PS

**9.** Plant cells and animal cells were left in water or 10% sucrose solution for 10 minutes. The cells were then examined under the microscope. The appearance of three individual cells is shown below.

Cell R | Cell S | Cell T

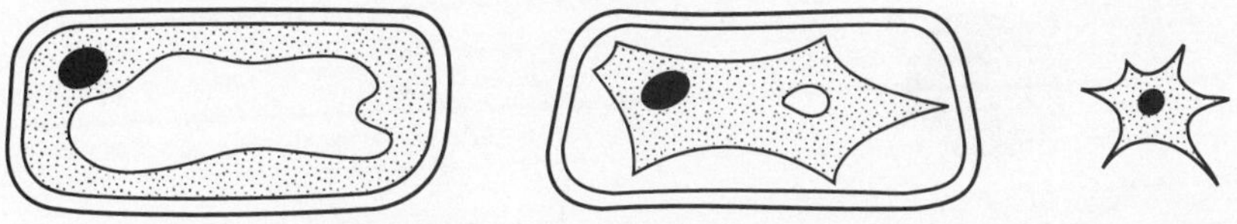

(*a*) Which **two** of the cells had been placed in 10% sucrose solution?

Cell _______ and Cell _______ **1**

(*b*) The change in the cells was caused by the movement of water into or out of the cells.

What is the name of this process?

____________________________ **1**

(*c*) With reference to the cells placed in water, what is meant by the term "concentration gradient"?

____________________________________________________________

____________________________________________________________ **1**

**[Turn over**

**10.** The diagram represents some of the stages of cell division.

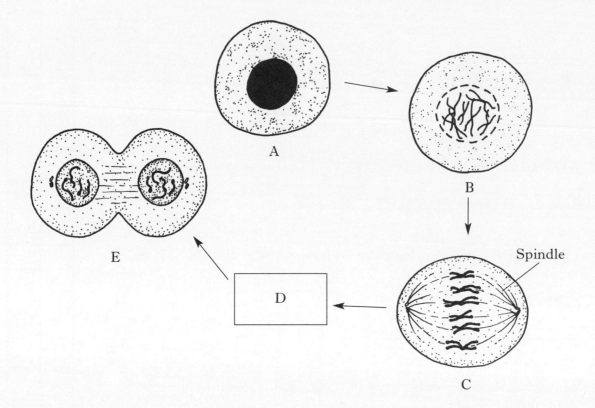

Describe what happens in stages C and D.

C _____

_____

D _____

_____

**11.** The activity of the enzymes lipase and catalase was investigated.

Three test tubes were set up.

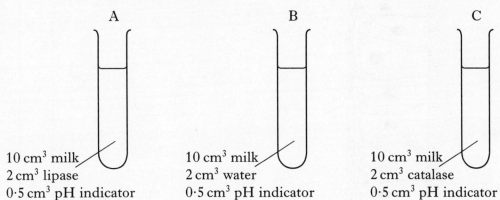

A: 10 cm³ milk, 2 cm³ lipase, 0·5 cm³ pH indicator
B: 10 cm³ milk, 2 cm³ water, 0·5 cm³ pH indicator
C: 10 cm³ milk, 2 cm³ catalase, 0·5 cm³ pH indicator

The colour of the pH indicator was noted at the start and after 20 minutes.
The results are shown in the table below.

| Test tube | Colour of pH indicator | |
|---|---|---|
| | At start | After 20 minutes |
| A | green | orange |
| B | green | green |
| C | green | green |

(a) In tube A, the pH indicator colour change was due to the production of fatty acids as the lipase reacted with the fat in the milk.

Explain why there was no change in tube C.

_____

_____ 1

(b) What term is used to describe tube B which contained water instead of an enzyme?

_____ 1

(c) Name **two** variables, not already shown, which would have to be kept the same when this investigation was set up.

1 _____

2 _____ 2

**12.** The diagram represents part of the breathing system in humans.

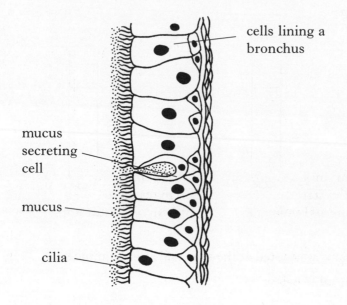

(a) Describe how the mucus and cilia help to protect the lungs from damage and infection.

Mucus _____

_____

Cilia _____

_____

(b) Which of the following are involved in **breathing out** during deep breathing in humans?

1 Diaphragm contracts
2 Diaphragm relaxes
3 Muscles between the ribs contract
4 Muscles between the ribs relax
5 Rib cage moves up and out
6 Rib cage moves down and in

*Tick the correct box.*

1, 3 and 5 correct ☐

1, 4 and 5 correct ☐

2, 3 and 6 correct ☐

2, 4 and 6 correct ☐

**12.** (continued)

(c) The pie charts below show the composition of fresh air and breathed air.

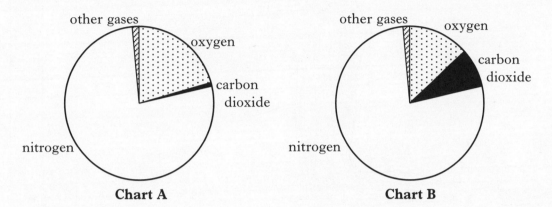

**Chart A**  **Chart B**

(i) Indicate which chart refers to breathed air and give a reason for your choice.

The chart that refers to breathed air is _____

Reason _____

_____ 1

(ii) Which named gas appears not to be involved in gas exchange in the lungs?

_____ 1

[Turn over

**13.** The diagram shows some of the structures in a human arm.

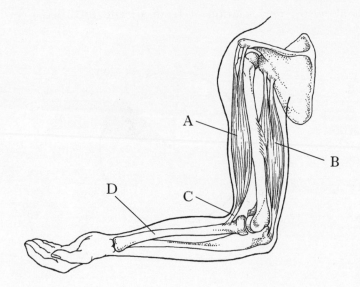

(a) Name the type of structures labelled A, B and C.

A and B _____

C _____        1

(b) Which structure contracts to bend the arm?

Letter _____        1

(c) Which of the following are composed of living cells?
*Tick the correct box.*

Structures A and B only ☐

Structures A, B and C only ☐

Structures A, B and D only ☐

Structures A, B, C and D ☐        1

**14.** The bar chart shows the blood flow to parts of the body when a person is sitting still.

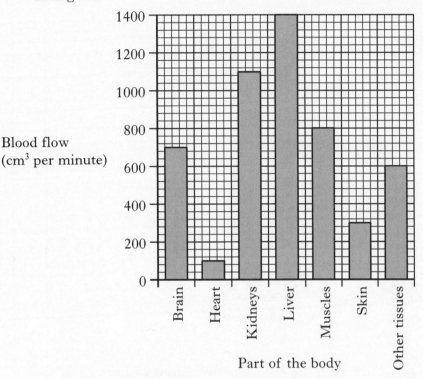

(a) What is the total blood flow per minute?
*Space for calculation*

Total _____ cm³

(b) Express the flow of blood to the liver, the brain and the heart as a simple whole number ratio.
*Space for calculation*

_____ : _____ : _____
liver    brain    heart

(c) During exercise the blood flow to the muscles increases to 1200 cm³ per minute.

Calculate the percentage increase in blood flow to the muscles.
*Space for calculation*

_____ %

**15.** (a) True breeding pea plants were bred to produce two generations, as shown below.

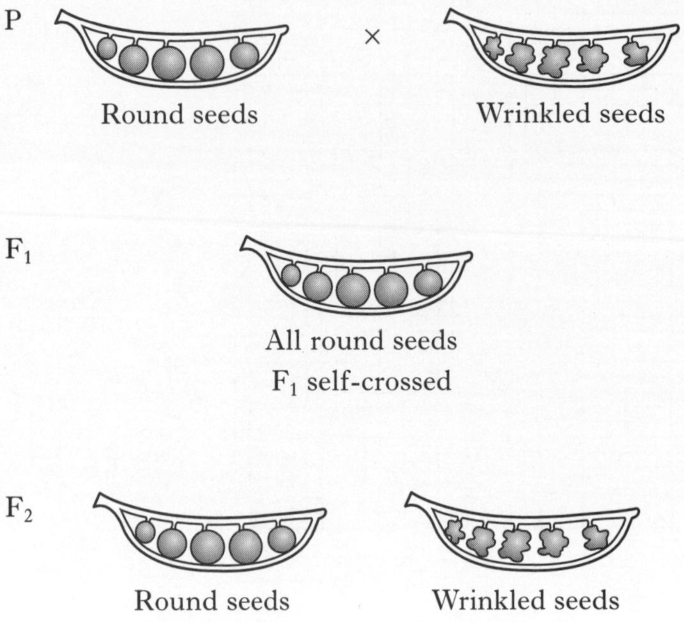

P  Round seeds  ×  Wrinkled seeds

F₁  All round seeds
F₁ self-crossed

F₂  Round seeds   Wrinkled seeds

(i) Round seeds and wrinkled seeds are caused by two forms of the same gene.

What term describes different forms of the same gene?

_____

(ii) Using the letter **R** for round seeds and **r** for wrinkled seeds, complete the following table.

| Plant | Genotype |
|---|---|
| Parent with round seeds | |
| All F₁ | |
| F₂ with wrinkled seeds | |

**15.** *(a)* **(continued)**

   (iii) The seeds from the $F_2$ were counted and the results are shown in the bar chart.

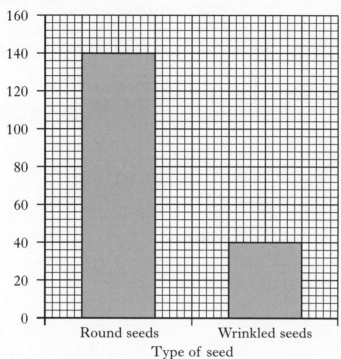

   What is the simple whole number ratio of round to wrinkled seeds?

   *Space for calculation*

   _____ : _____
   round    wrinkled

*(b)* In order to improve a certain characteristic, particular pea plants are chosen to breed.

   (i) What is the name given to this procedure?

   _____

   (ii) Suggest a characteristic of pea plants which a plant breeder might wish to improve.

   _____

[Turn over

**16.** (a) Exposure to radiation can cause mutation.
The pie chart shows the contribution of various sources of radiation to the total exposure.

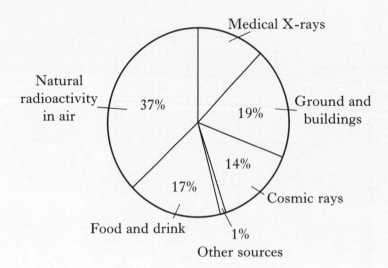

(i) Which source of radiation contributes most to the total exposure?

_____

(ii) What percentage of the total exposure comes from X-rays?
*Space for calculation*




_____ %

**16. (continued)**

(b) The table shows the occurrence of chromosome mutations in *Drosophila* fruit flies when exposed to different doses of radiation.

| Dosage of X-rays (millisieverts) | Chromosome mutations (%) |
|---|---|
| 1000 | 1·0 |
| 2000 | 1·9 |
| 2500 | 2·6 |
| 3000 | 3·1 |
| 4000 | 4·2 |
| 4500 | 4·6 |
| 5000 | 5·3 |

(i) On the grid below, complete the y-axis and plot a line graph of the results.

(An additional grid, if needed, will be found on page 27.)

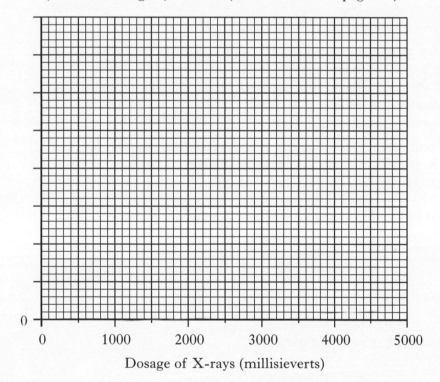

(ii) What is the relationship between the dosage of X-rays and chromosome mutations?

_____

_____

**17.** Read the following passage and answer the questions based on it.

Midges are blood-feeding insects similar to mosquitoes but much smaller. There are about 40 species found in Scotland. One species, *Culicoides impunctatus* (or the 'Highland biting midge'), is best known for biting people.

Midges inhabit all areas of the world except the poles, New Zealand and Patagonia (where the habitat is too dry). In the UK, their numbers are greatest in parts of Western Scotland and the Highlands where they thrive in the damp acidic soil. The colder winters on the East coast often result in frozen soil which kills the overwintering larvae.

In late summer, eggs are laid on the soil surface. They hatch into thread-like larvae which live a few centimetres below the soil surface and feed on decaying plant and animal matter. Flying adults begin to emerge the following May, triggered by lengthening days and warmer temperatures. These adults lay eggs that develop quickly to give a second emergence of adult midges in August.

Midges are unable to fly in very wet or windy weather or in temperatures below 7 °C. In such poor conditions they might only survive a few days. In more favourable conditions they may live for two weeks.

All midges feed on plant nectar. Only females feed on blood as they require the blood proteins and fats to develop their eggs.

Different species of midge specialise in feeding on different hosts. The Highland biting midge feeds on large mammals, including cattle, horses, deer, and of course, people. The host is detected by a combination of smells, heat, carbon dioxide, movement and colour. Differences amongst people in these factors partly explain why midges bite some more than others.

Biting midges and mosquitoes obtain blood in different ways. Mosquitoes insert their mouth-parts directly into a blood capillary. Midges use their jaws to cut a hole in the skin, creating a pool of blood from which they feed.

(a) Give **two** reasons why midges do not exist in some parts of the world.

1 _____

2 _____

(b) What do the midge larvae feed on?

_____

(c) What is the maximum time an adult midge can live for?

_____

DO NOT WRITE IN THIS MARGIN

Marks KU PS

**17. (continued)**

(*d*) Name **two** types of food that an adult female midge might eat.

1 ______________________________

2 ______________________________ **1**

(*e*) Suggest why the Highland biting midge might not bite mice.

______________________________

______________________________ **1**

(*f*) Describe **two** differences, mentioned in the passage, between midges and mosquitoes.

1 ______________________________

2 ______________________________ **1**

**[Turn over**

**18.** An investigation into fermentation was carried out at 20 °C.

A deflated balloon was attached to the top of each tube at the start. The appearance of the balloons after several hours is shown below.

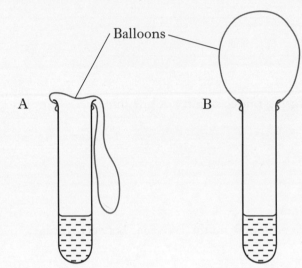

Boiled yeast suspension and sugar solution

Yeast suspension and sugar solution

(a) What substance produced in tube B caused the balloon to inflate?

_____

(b) The balloon on the tube with the boiled yeast suspension did not inflate. Explain this result.

_____

_____

(c) How would the appearance of the balloon on tube B differ if the investigation had been carried out at 15 °C?

_____

**19.** The diagrams show the production of insulin by genetic engineering. They are not in the correct order.

**A** Plasmids removed from bacteria

**B** Insulin gene identified

**C** Plasmids combined with insulin gene

**D** Altered plasmids mixed with suitable bacteria

**E** Genetically engineered bacteria produce insulin

**F** Enzymes cut the insulin gene from the chromosome

(a) Use the letters to put the diagrams in the correct sequence.
The first and last have been done for you.

B ☐ ☐ ☐ ☐ E

(b) Explain why there is an ever-increasing need for insulin produced by genetic engineering.

_____

_____

(c) Before biotechnology was used to produce insulin, it was obtained from the pancreas of animals such as pigs.

Give **one** advantage of producing insulin by genetic engineering.

_____

_____

[Turn over

DO NOT WRITE IN THIS MARGIN

Marks KU PS

**20.** Four test tubes were set up to investigate decay of beetroot as shown below.

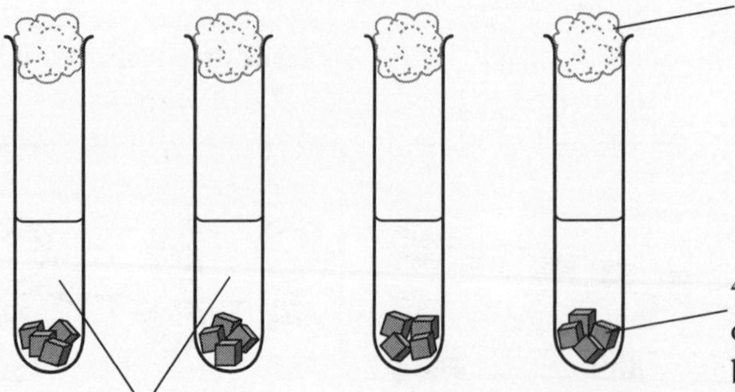

Test tube A was incubated at 5 °C and tubes B, C and D were incubated at 30 °C. After 48 hours, the appearance of the solutions was recorded. Any cloudiness in the solution was due to the growth of micro-organisms.

The results are shown in the table.

| *Test tube* | A | B | C | D |
|---|---|---|---|---|
| *Appearance* | Clear | Cloudy | Slightly cloudy | Clear |

(*a*) (i) In which test tube would the beetroot show most decay if left for two weeks?

Test tube ____________ **1**

(ii) From the results, suggest **two** methods of preventing decay in beetroot.

1 ______________________________________________

2 ______________________________________________ **1**

(iii) Suggest why cotton wool plugs were used in the experiment.

______________________________________________

______________________________________________ **1**

(*b*) Name a type of micro-organism responsible for decay.

______________________________________________

______________________________________________ **1**

[*END OF QUESTION PAPER*]

ADDITIONAL GRAPH PAPER FOR QUESTION 16(b)(i)

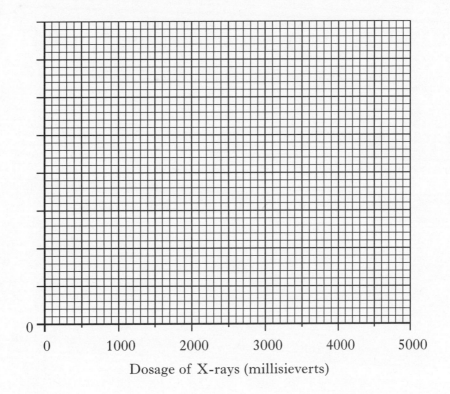

[Turn over

SPACE FOR ANSWERS
AND FOR ROUGH WORKING

**2005** | Credit

# 0300/402

**FOR OFFICIAL USE**

**C**

KU | PS

Total Marks

NATIONAL QUALIFICATIONS 2005

WEDNESDAY, 18 MAY 10.50 AM – 12.20 PM

**BIOLOGY STANDARD GRADE**
Credit Level

---

**Fill in these boxes and read what is printed below.**

Full name of centre

Town

Forename(s)

Surname

Date of birth
Day Month Year

Scottish candidate number

Number of seat

1. All questions should be attempted.

2. The questions may be answered in any order but all answers are to be written in the spaces provided in this answer book, and must be written clearly and legibly in ink.

3. Rough work, if any should be necessary, as well as the fair copy, is to be written in this book. Additional spaces for answers and for rough work will be found at the end of the book. Rough work should be scored through when the fair copy has been written.

4. Before leaving the examination room you must give this book to the invigilator. If you do not, you may lose all the marks for this paper.

**1.** (*a*) Use information from the diagrams of invertebrates to complete the following paired statement key.

The diagrams are not to the same scale.

spider  
dragon fly  
flea  
human louse  
mite  
house fly

1  Wings present .................................................................. Go to ☐   **1**

   Wings absent .................................................................. Go to 3

2  One pair of wings ................................................................ house fly

   ☐ .................................................................. dragon fly   **1**

3  Four pairs of legs .............................................................. Go to 4

   Three pairs of legs ........................................................... Go to 5

4  Body clearly divided into two parts ................. ☐

   Body not clearly divided into two parts ........... ☐   **1**

5  Hooked claw on legs ....................................................... human louse

   No claw on legs .................................................................. flea

(*b*) Give **two** features mentioned in the key which the human louse and the flea have in common.

1 _____

2 _____   **1**

2. The light intensity inside and outside a woodland was measured over a year. The table shows the results.

|  | Average daily light intensity (units) | |
| --- | --- | --- |
| Month | Outside woodland | Inside woodland |
| January | 10 | 8 |
| February | 13 | 10 |
| March | 15 | 12 |
| April | 19 | 16 |
| May | 24 | 22 |
| June | 28 | 15 |
| July | 30 | 5 |
| August | 25 | 5 |
| September | 20 | 5 |
| October | 15 | 5 |
| November | 12 | 10 |
| December | 10 | 8 |

(a) On the grid below, complete the Y axis, key and line graph plot to show the results.
(An additional grid, if required, will be found on page 22.)

**Key**

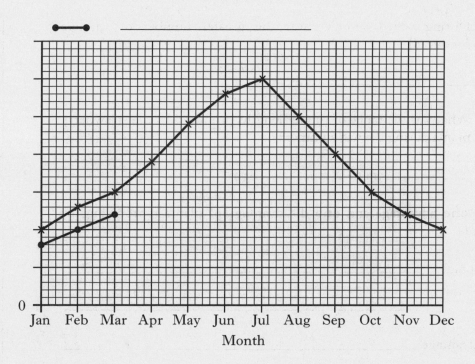

(b) Explain why the difference between the light intensities inside and outside the woodland is much greater from June to October.

_____

**3.** Strawberry plants reproduce by runners. The bar chart below shows the number of runners of different lengths produced by the same plants during a dry summer and a wet summer.

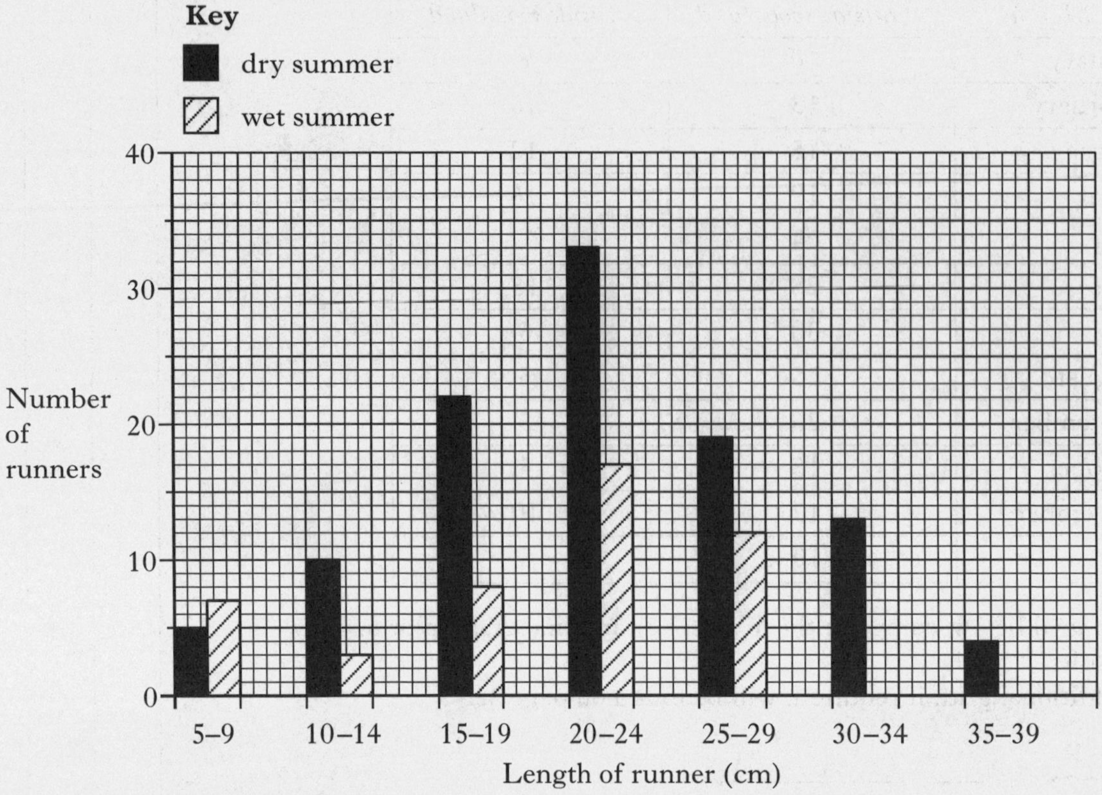

**Key**
- ■ dry summer
- ▨ wet summer

(a) (i) During which summer was the greater number of runners produced?

_____

(ii) What was the range of runner lengths produced by the strawberry plants during the wet summer?

_____ cm to _____ cm

(b) State **one** advantage and **one** disadvantage of plants reproducing by runners.

Advantage _____

_____

Disadvantage _____

_____

(c) What word is used to describe the type of reproduction of which runners is an example?

_____

3. **(continued)**

(d) Plants may also produce seeds which can be dispersed away from the parent plant. The diagrams below show some seeds and fruits of named plants. They all use one of two methods of seed dispersal.

sycamore

dandelion

goosegrass

burdock

ash

(i) Complete the following table to identify each of these methods of seed dispersal and the plants which use them.

| *Method of seed dispersal* | *Plants which use this method* |
|---|---|
|  |  |
|  |  |

(ii) Plants with succulent fruits use a different method of seed dispersal.

Describe this method.

_____

_____

[Turn over

4. In an investigation into the rate of photosynthesis, a piece of *Elodea* (pondweed) was placed in a beaker of water and a bright light shone on it.

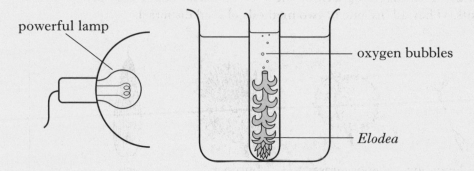

Bubbles of oxygen given off from the *Elodea* were counted. This was repeated with the lamp at different distances from the *Elodea*.

A graph of the results is shown below.

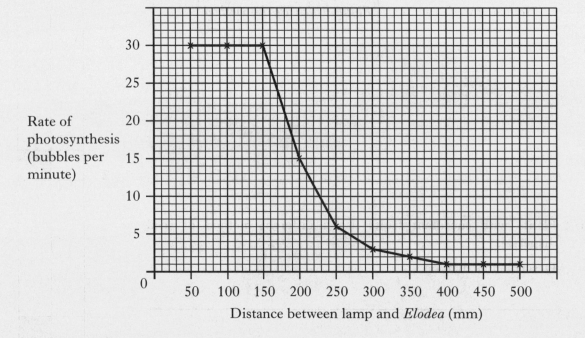

(a) (i) What was the rate of bubbling when the lamp was at a distance of 200 mm from the *Elodea*?

_____ bubbles per minute

(ii) What distance between the lamp and the *Elodea* would give a rate of bubbling of 5 bubbles per minute?

_____ mm

(iii) What would happen to the rate of bubbling if the container with the *Elodea* was placed in the dark?

_____

**4. (continued)**

(b) (i) At which distances between the lamp and the *Elodea* did light act as a limiting factor on the rate of photosynthesis?
*Tick the correct box.*

50 – 150 mm ☐

150 – 400 mm ☐

400 – 500 mm ☐

(ii) Name **one** other factor which could limit the rate of photosynthesis.

_____

(c) The investigation was carried out several times and the average results were used to plot the graph. Why was this good experimental technique?

_____

[Turn over

5. The water quality at beaches is tested to check that it is not affected by any untreated sewage.

The table gives information about the number of beaches which were tested in one particular year and the number passed as suitable for swimming.

| Country | Number of beaches tested | Number of beaches suitable for swimming | Percentage of beaches suitable for swimming |
|---|---|---|---|
| England | 271 | 239 | 88·2 |
| Scotland | 93 | 68 | |
| Wales | 128 | 102 | 79·7 |
| Northern Ireland | 17 | 16 | 94·1 |

(a) Complete the table to show the percentage of beaches suitable for swimming in Scotland.

*Space for calculation*

(b) Why should the percentages of beaches which passed be used when comparing the results from the four countries, rather than the actual number?

(c) The samples of water from the beaches can be examined for the presence of certain species. This technique gives information about water pollution. What name is given to such species?

6. The diagram below shows part of the nitrogen cycle.

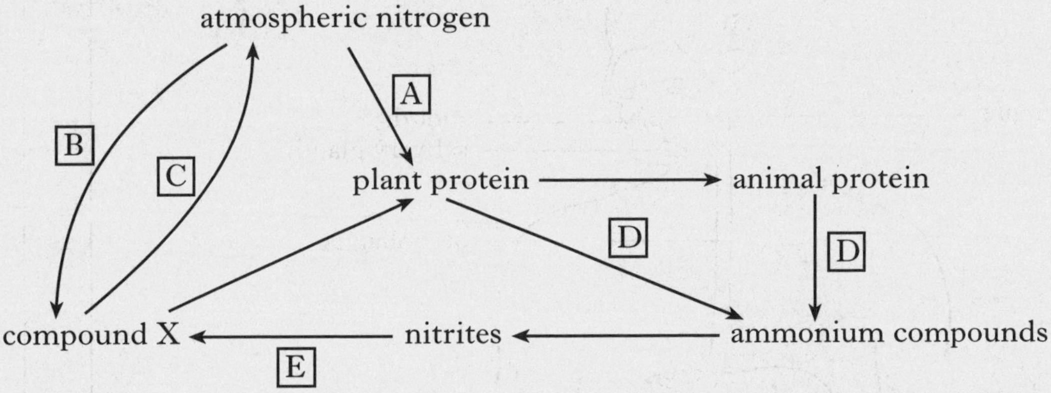

(a) Use letters from the diagram to complete the following table about some of the events of the nitrogen cycle.

| Event | Letter |
|---|---|
| Death and decay | |
| Action by denitrifying bacteria | |
| Lightning | |

2

(b) Explain why event A can take place in some plants such as clover, peas and beans, but not in others.

_____

_____

1

(c) Name compound X.

_____

1

[Turn over

**7.** The following diagram shows the human digestive system.

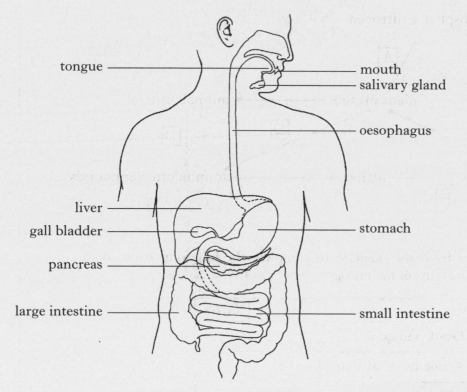

(a) Name **three** structures from the diagram that produce digestive juices.

1 _____

2 _____

3 _____

(b) (i) Complete the following table to show the substrates and products of some digestive enzymes.

| Enzyme | Substrate | Product(s) |
|---|---|---|
|  |  | maltose |
|  | protein |  |
| lipase |  | 1 ..................... <br> 2 ..................... |

(ii) What word describes the conditions in which enzymes work best?

_____

**8.** The following graph shows the results of an investigation on the effect of ADH on urine production.

Line A shows the rate of urine production for a volunteer after drinking one litre of water.

Line B shows the rate of urine production from the same volunteer after drinking one litre of water and receiving an injection of ADH.

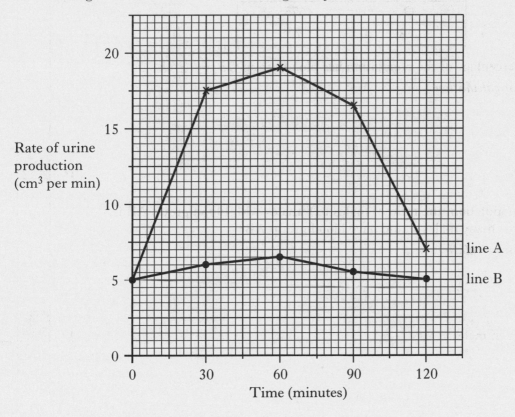

(a) (i) What effect did ADH have on the production of urine?

_____

(ii) How long did it take for the rate of urine production to reach its maximum without an injection of ADH?

_____ minutes

(iii) After 15 minutes, what was the difference between the rates of urine production with and without the ADH injection?

*Space for calculation*

_____ cm³ per minute

(b) Which organs of the body respond to the presence of ADH?

_____

**9.** The table shows the number of people with each blood group in a population of 1500.

| Blood group | Number of people |
|---|---|
| A | 610 |
| B | 143 |
| O | 675 |
| AB | 72 |

(a) What percentage of the population has blood group O?

*Space for calculation*

_____ %

(b) In the population, the ratio of males to females with blood group AB is 5:3. How many males would have blood group AB?

*Space for calculation*

Number of males _____

9. **(continued)**

(c) Blood platelets are important in the formation of blood clots at the site of an injury. The following diagram shows the sequence of reactions which produce the clot when platelets gather at the injury.

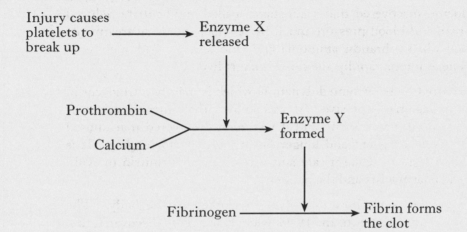

(i) Suggest **one** benefit of a blood clot forming at the site of an injury.

_____ 1

(ii) Explain why low calcium levels would reduce the blood's ability to clot.

_____
_____ 1

**[Turn over**

**10.** Read the following passage and answer the questions based on it.

**An Explosive Medication**

Angina is a pain in the chest that happens when the heart muscle does not receive enough blood. It occurs when branches of the artery that carries blood to the heart become narrowed or blocked. In 1867, T. L. Brunton, an Edinburgh doctor, discovered that a substance called amyl nitrite reduced both angina pain and blood pressure much better than the usual treatments which included whisky, brandy, ammonia and chloroform. Although amyl nitrite does relieve angina rapidly, the relief is short lived.

Also in 1867, Alfred Nobel invented dynamite which is mainly nitroglycerin, a powerful and unstable explosive. It was known that nitroglycerin was similar in structure to amyl nitrite and it was soon discovered that diluted nitroglycerin was an excellent and longer lasting remedy for angina. It is used in a diluted form to make it safe and was renamed Trinitrin to avoid scaring both the pharmacists and the patients.

However, it was a mystery how nitroglycerin worked in the body. The mystery remained unsolved until the 1970s when researchers discovered that it works by changing into nitric oxide. Outside the body, nitric oxide is a poisonous gas but it plays a vital part inside the body. Nitric oxide is the main messenger making blood vessels open wider so that more blood flows to the starved heart muscle, and this is why nitroglycerin helps angina patients.

(a) What is the name of the artery which carries blood to the heart muscle?

_____

(b) What is the cause of the chest pain referred to as angina?

_____

(c) What made nitroglycerin appear suitable for investigating as a possible heart medicine?

_____

(d) Nitroglycerin is diluted to make it safe. Why is this necessary?

_____

(e) Complete the following flow diagram to show how nitroglycerin works in the body.

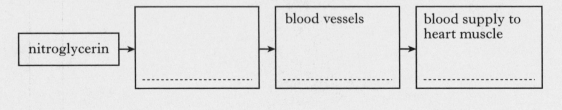

**11.** The diagrams below show some of the muscles in the leg.

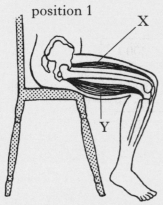
position 1

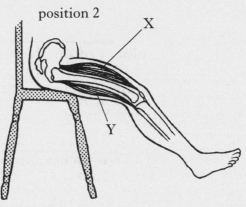
position 2

*(a)* (i) Describe the action of muscles X and Y as the leg moves from position 1 to position 2.

Muscle X _____

Muscle Y _____     1

(ii) Name the structures which attach muscles to bones.

_____     1

*(b)* (i) Draw a line from each of the following parts of the brain to its correct function.

| Part | Function |
|---|---|
| cerebrum | controls heart rate |
| cerebellum | controls balance |
| medulla | enables conscious thought and memory |

2

(ii) Underline **one** option in each bracket to describe the flow of information in the nervous system.

Information from the environment is detected by the { heart / sense organs / brain }

and sent to the { central nervous system / circulation system / skin } which responds by

sending messages to the { muscles / blood / bones }.     1

**12.** The table gives the average yield of maize (corn) per hectare in the USA since 1960.

| Year | 1960 | 1970 | 1980 | 1990 | 2000 |
|---|---|---|---|---|---|
| Average annual yield (tonnes per hectare) | 4·05 | 5·32 | 7·14 | 8·21 | 10·05 |

(a) (i) Calculate the average annual increase in yield in the 40 year period from 1960 to 2000.

*Space for calculation*

Average annual increase in yield _____ tonnes per hectare

(ii) Which ten year period showed the greatest average increase in yield?

Tick the correct box.

*Space for calculation*

1960 – 1970 ☐    1980 – 1990 ☐

1970 – 1980 ☐    1990 – 2000 ☐

(b) Each individual plant in the field gives a different yield which can be any value between the lowest and the highest.

What name is given to this type of variation?

_____

**12. (continued)**

(c) (i) The improvement in yield has been largely due to the production of new varieties of maize by *selective breeding*.

Explain what is meant by this term.

_____

_____

_____

(ii) It is possible to produce new varieties of maize by introducing genes from species which do not interbreed with maize.

What general name is given to these techniques?

_____

(d) In order to increase the variation available for selective breeding, plant biologists treat maize in ways that can increase the rate of mutation.

(i) What is meant by the term *mutation*?

_____

(ii) Give an example of a factor that can increase the rate of mutation.

_____

(e) Farmers try to ensure the maximum yield of crops. This requires a plentiful supply of plant nutrients and little competition from other plants.

Describe how each of these can be achieved.

Plentiful supply of plant nutrients _____

_____

Reduced competition from other plants _____

_____

[Turn over

**13.** The table shows a comparison of the breakdown of one gram of glucose by three different types of cell respiration.

|  | Type of cell respiration | | |
|---|---|---|---|
|  | A | B | C |
| Energy released (kJ) | 17·1 | 0·9 | 0·9 |
| Oxygen used (g) | 1·07 | 0 | 0 |
| Carbon dioxide produced (g) | 1·47 | 0 | 0·49 |
| Water produced (g) | 0·6 | 0 | 0 |
| Lactic acid produced (g) | 0 | 1 | 0 |
| Ethanol produced (g) | 0 | 0 | 0·51 |

(a) (i) Respiration of type A releases much more energy than the other types.

What name is given to this type of respiration?

_____

(ii) Which **two** types of respiration take place in the following cells?

Muscle cells:    type _____ and type _____

Yeast cells:    type _____ and type _____

(b) Express the energy released from one gram of glucose by the three types of respiration as a simple whole number ratio.

*Space for calculation*

_____ : _____ : _____
type A   type B   type C

(c) Give **one** way in which the chemical energy released from food is important in the metabolism of cells.

_____

_____

**14.** In an investigation on gas exchange, samples of breathed air were collected from several volunteers. The table shows the volumes of carbon dioxide and oxygen in 1000 cm³ of each sample.

| Sample | Volume of carbon dioxide (cm³) | Volume of oxygen (cm³) |
|---|---|---|
| A | 10 | 153 |
| B | 7 | 148 |
| C | 6 | 154 |
| D | 11 | 153 |
| E | 6 | 152 |
| Average | 8 | |

(a) Complete the table by calculating the average volume of oxygen in the samples.
*Space for calculation*

(b) Calculate the percentage of oxygen in sample C.
*Space for calculation*

_____ %

(c) Name the chemical in the blood which combines with oxygen to transport it to the body tissues.

_____

[Turn over

**15.** The table shows the composition of 100 g of four common fruits.

| Fruit | Component | | | |
|---|---|---|---|---|
| | Protein (g) | Carbohydrate (g) | Fat (g) | Water (g) |
| bananas | 1·0 | 23 | 0·3 | |
| apples | 0·4 | 12 | 0·1 | 87·5 |
| pears | 0·4 | 10 | 0·1 | 89·5 |
| grapes | 0·2 | 15 | 0·1 | |

(a) Complete the table by adding the mass of water present in 100 g of bananas and grapes.

*Space for calculation*

(b) Which component of the fruits contains the most energy per gram?

_____

(c) Give **two** differences between the composition of apples and that of pears.

1 _____

2 _____

(d) Suggest how it would be possible to minimise the effect of variations of individual fruits when measuring their composition.

_____

_____

**16.** Identical pieces of cloth were marked with stains. They were then washed at different temperatures using biological detergent. The degree of staining still on the cloth after washing was measured and expressed as a percentage of the stain before washing. The test was repeated using a non-biological detergent.

The results are shown in the graph below.

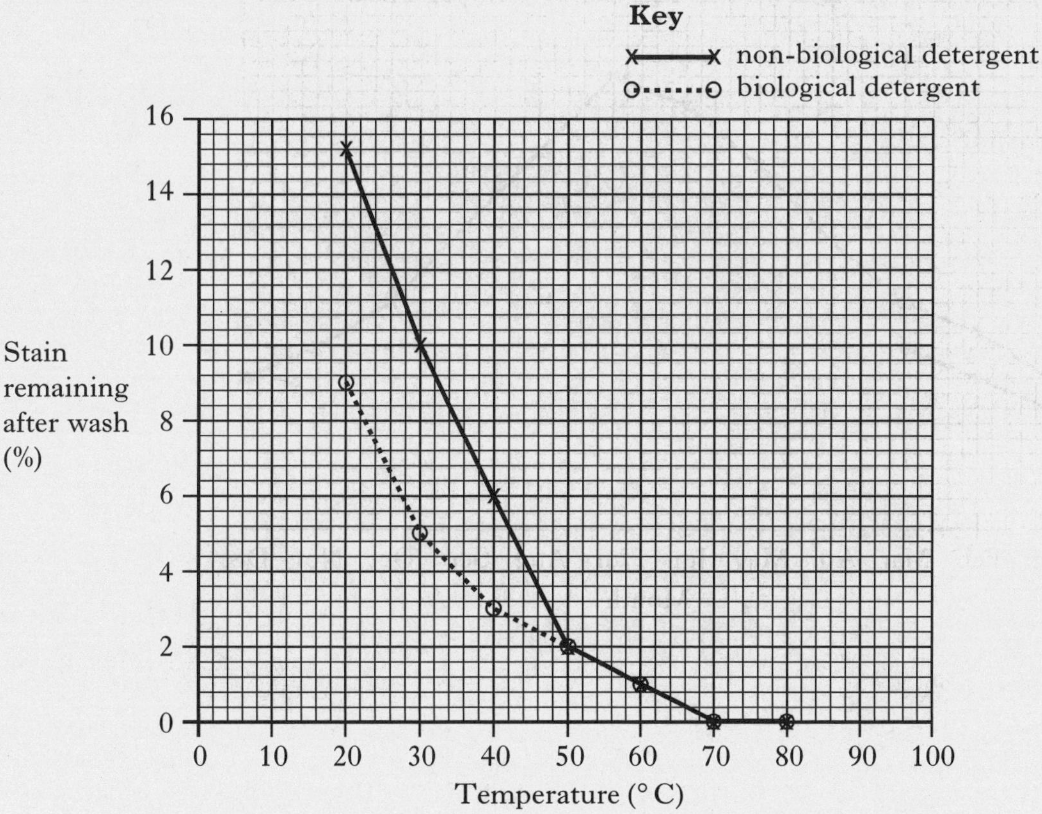

**Key**
×———× non-biological detergent
o·······o biological detergent

*(a)* (i) Suggest **one** precaution concerning the stains which would be necessary to ensure that the experimental procedure was valid.

_____

(ii) Compare stain removal by the two detergents over the temperature range shown.

_____

_____

*(b)* (i) Explain the action of biological detergents.

_____

(ii) Explain the economic advantage of using a biological detergent.

_____

_____

*[END OF QUESTION PAPER]*

ADDITIONAL GRAPH PAPER FOR QUESTION 2(a)

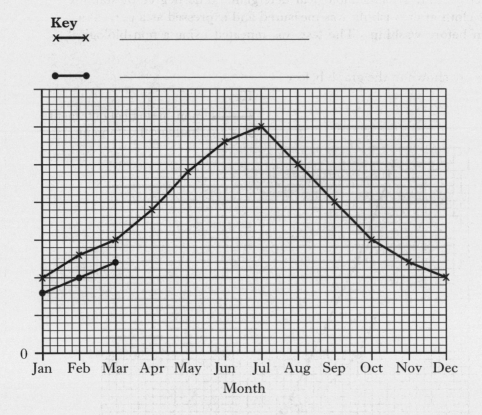

SPACE FOR ANSWERS
AND FOR ROUGH WORKING

SPACE FOR ANSWERS
AND FOR ROUGH WORKING

[BLANK PAGE]

[BLANK PAGE]

# Acknowledgements

Leckie & Leckie is grateful to the copyright holders, as credited, for permission to use their material:
Phillip Allen Publishers for the article 'All The Better To See You With' from *Biological Sciences Review* Vol 8 (November 1995) (2002 paper p 18).

The following companies have very generously given permission to reproduce their copyright material free of charge:
Mitchell Beazley Publishers for the article 'Great Oaks from Little Acorns Grow', adapted from the *Royal Horticultural Society's Encyclopaedia of Practical Gardening* (2003 paper p 16).